COKE OF NORFOLK

Coke with shepherd and sheep
(*From an engraving by William Ward*)

Coke of Norfolk

A FINANCIAL AND
AGRICULTURAL STUDY
1707–1842

BY

R. A. C. PARKER

FELLOW OF THE QUEEN'S COLLEGE
OXFORD

OXFORD
AT THE CLARENDON PRESS
1975

Oxford University Press, Ely House, London W. 1

GLASGOW NEW YORK TORONTO MELBOURNE WELLINGTON
CAPE TOWN IBADAN NAIROBI DAR ES SALAAM LUSAKA ADDIS ABABA
DELHI BOMBAY CALCUTTA MADRAS KARACHI LAHORE DACCA
KUALA LUMPUR SINGAPORE HONG KONG TOKYO

ISBN 0 19 822403 6

© Oxford University Press 1975

All rights reserved. No part of this publication may be reproduced, stored in a retrieval system, or transmitted, in any form or by any means, electronic, mechanical, photocopying, recording, or otherwise, without the prior permission of Oxford University Press

*Printed in Great Britain
at the University Press, Oxford
by Vivian Ridler
Printer to the University*

PREFACE

THIS book is intended as a contribution to knowledge and understanding of the economic activities of landlords in England in the eighteenth and early nineteenth centuries. It deals with one eminent and famous example, Thomas William Coke of Holkham, and seeks to elucidate the contribution he and his advisers made to the economic development of his estates. I have not attempted to compare the history of this estate with that of others in the same period since this would have involved the rewriting of a large part of the history of eighteenth- and early nineteenth-century agriculture.

It is evident, however, that the actions of a landlord cannot satisfactorily be explained without an examination of his wider financial and economic circumstances—in particular, his non-agricultural sources of money and his other outlets for spending, especially on the upkeep and management of a great house and on provision for families. Thus chapters of this book deal with general finance, and special sections are devoted to the Dungeness lighthouse, which brought about the transfer of money from shipowners to the Cokes. Moreover, the work of Coke of Norfolk cannot be assessed without a study of the economic circumstances he inherited. Hence the first part of the book covers the years between 1707 and the inheritance of his estates by Thomas William Coke in 1776.

In 1707, when the estates were inherited by Thomas Coke, later Lord Lovell, and subsequently earl of Leicester, as a minor, the series of audit account books, the major source for a history of the estate, began. In 1718 Thomas Coke came of age. By the time of his death in 1759, major agricultural advances had taken place and the house at Holkham was almost complete. A description of his activities and circumstances is, therefore, an essential preliminary to a discussion of Coke of Norfolk's career. Unfortunately very little correspondence survives from that period so that much reliance has had to be placed on analysis of figures, a fact which causes the first part of the book to be more austere and impersonal in content than might be wished. From 1759 to 1775 the estate was controlled by the widowed Lady Leicester and then, for only one year, by Wenman Coke (formerly Roberts), Lord Leicester's nephew. Coke of

Norfolk was his son. By the time of Coke's death in 1842, his property had become concentrated in Norfolk, and it is his activity in that county that is significant, so that I have devoted attention almost exclusively to the Coke estates in Norfolk, the largest exception being the Lancashire properties which offer some unusual features of interest.

This book is based on the manuscript records preserved at Holkham. I am most grateful to the present earl of Leicester, Coke's great-great-great grandson, for enabling me to work on them there. (Some of the main sources are now available on microfilm at the Bodleian Library in Oxford.) The staff of the Holkham estate office treated me with patience and courtesy. I must thank, too: Dr. W. O. Hassall, who acts as librarian at Holkham; Mr. S. Powell, of the Oxford University Institute of Statistics for his help in constructing the moving averages shown in the graph (p. 212); and Miss Pat Lloyd, who drew the graph. Finally, I am much indebted for help and advice to Mr. H. J. Habakkuk, Principal of Jesus College, Oxford; Mr. J. P. Cooper of Trinity College, Oxford; Mrs. Janet Howarth of St. Hilda's College, Oxford; Mr. C. M. Edwards of Queen's College, Oxford; and to Mrs. Jennifer Loach of Somerville College, Oxford.

CONTENTS

Coke with Shepherd and Sheep   *Frontispiece*

TABLES   ix

MAPS AND PLANS   x

ABBREVIATIONS   xi

NOTE ON SOURCES   xiii

1. The Estates and Finances under the Guardianship, 1707–1718   1

2. Thomas Coke and the South Sea Bubble   12

3. General Finance, 1722–1759: the Building of Holkham Hall and the Exploitation of the Lighthouse   21

4. The Estates under Thomas Coke, first earl of Leicester   37
    i. The size and value of the estates   37
    ii. The improvement of the estates   39
    iii. Investment in, and administration of, the estates   53
    iv. The home farm   57

5. The Estate and Finances under Margaret, countess of Leicester, and Wenman Coke, 1759–1776   61

6. Coke of Norfolk and Agricultural History   71

7. The Estate under Coke, 1776–1816   83
    i. The structure of ownership and tenancy   83
    ii. Investment, improvement, and growth in rent   93
    iii. Leases and husbandry covenants   100
    iv. Crop rotations employed by tenants   105

8. The Park farm and the Holkham sheep-shearings, 1776–1821   114

Contents

9. General Finance, 1776-1822 — 126

10. The Estate under Coke, 1816-1842 — 135
    i. Francis Blaikie, Steward — 135
    ii. Leases and their enforcement — 138
    iii. The crisis, 1821-1822 — 147
    iv. Investment in the estates and the coming of 'high farming' — 152
    v. Tenants' cropping and size of farms — 161
    vi. The condition of labourers — 165

11. The Park farm—the Post-War Crisis and after — 169

12. General Finance, 1822-1842 — 175
    i. The Lancashire estate and Coke's coal-mine — 175
    ii. The loss of the Dungeness lighthouse tolls — 178
    iii. The crisis of 1822 and the recovery — 188

13. Some Conclusions — 199

APPENDICES

1. The management of the lighthouse — 202
2. A note on the work 'Break' — 204
3. Calculations, *c*. 1815 about Longham Hall farm — 205
4. 1822: Expenses considered as belonging to the domestic establishment — 206
5. A note on prices — 207

Graph — 212

Chart of descent — 213

Index — 215

TABLES

A.	Rents 1629, 1651, 1656, 1667, 1677, 1706	4
B.	Receipts for the Home Farm, 1750-3	58
C.	Disbursements for the Home Farm, 1750-3	59
D.	Size of Farms in 1780	89
E.	Cropping at Flitcham, 1789-91 and 1800-2	107
F.	Cropping at Longham, 1789-91 and 1800-2	108
G.	Cropping at Massingham, 1789-91 and 1800-2	109
H.	Cropping at Massingham, 1789-1802	110
I.	Total crops grown on the estates, 1790-7	113
J.	Crops grown on the Park Farm, 1782-7	124
K.	Yields per acre on the Park Farm, 1782-7	124
L.	Grain produced on the Park Farm, 1782-7	125
M.	Norfolk Rents and Arrears, 1814-24	146
N.	Norfolk Rents, 1820-42	153
O.	Cropping at Holkham, 1789-91 and 1851-3	162
P.	Cropping at Castleacre, 1789-91 and 1851-3	163
Q.	Size of Farms, 1780 and 1851	164
R.	Gross and Net Rents, whole estate, 1821-42	194

MAPS AND PLANS

Castleacre	44
North-western corner of Tittleshall	47
Harpley Dam Farm	47
Northern area of Longham	98
Map of Norfolk	214

ABBREVIATIONS

A/B	Audit Book(s)
A.L.B.	Agricultural Letter Books (2 vols.)
G.E.D.	General Estate Deeds
H.F.D.	Holkham Family Deeds
J.R.A.S.E.	*Journal of the Royal Agricultural Society of England*
Keary	Report on the Holkham estates by E. Keary (2 vols., 1851)
L.B.	Letter Book
Norwich P.L.	Norwich Public Library (which contains most of the documents, originally at Holkham, relating to the Flitcham estate which was sold to King George V)
R.A.S.E.	Royal Agricultural Society of England

NOTE ON SOURCES

THIS study is based on records preserved at Holkham. Some of them were in the muniment room next to the estate office, some in the estate office, some in the powder room above the game larder, some in the manuscript section of the library. The location of documents referred to in the text has been given except when the documents were in the muniment room. Since I worked on these records, two major changes have taken place. Some of the documents have been microfilmed and copies of the films deposited in the Bodleian library in Oxford. At Holkham, the former contents of the game larder are now stored elsewhere.

The muniment room contained the series of audit account books starting from 1707 and continuing beyond 1842 in which there are gaps only for the years 1719-21 and 1759-74. It also contained the estate plans on racks in numbered order and the large volume of estate maps of 1779 referred to in the text as 'Plans' from the title on its cover. The game larder contained the series of letter books which start in 1816 and also various account books removed from the muniment room. The estate office contained a series of survey books distinguished in the text by the titles on the cover—two have the title 'Plans and Particulars of Norfolk Estate', and one of these is distinguished in the text by the addition '(Dugmore)' for 'Plans and Particulars of such Estates in Norfolk belonging to Thos. Willm. Coke Esq. as were surveyed by Jno. Dugmore'. It also contained Keary's report of 1851 and the letter-book for 1816. The library contained most of the other surviving letters. There are very few letters from or to Holkham until the letter books begin in 1816; this is the most serious defect in the sources, a defect which has unfortunately often made it impossible to explain the motives of actions whose consequences are fully recorded.

For the most part, the documents preserved in the muniment room have been arranged and listed. There are three series of documents thus ordered and numbered: the 'Holkham Family Deeds' relating to general family affairs, the 'General Estate Deeds' relating to general estate concerns, and the deeds, etc., relating to individual divisions of the estate, listed separately for each division, as, for example, 'Castleacre Deeds'.

1

The Estates and Finances under the Guardianship, 1707-1718

IN 1707 the Coke estates were inherited by a boy of ten, Thomas Coke. His estates lay in Norfolk, Suffolk, Buckinghamshire, Oxfordshire, Somerset, and London, and brought in annual rents of about £5,800, supplemented each year by casual profits, mainly from fines paid for renewals of long leases and sales of wood, of up to £1,000.[1] There were also lands, which would in due course revert to the estate, then held by Thomas's grandmother as her jointure, lands which were in Staffordshire, Kent, Dorset, and Suffolk. Thomas's mother, Cary Coke, died in 1707, soon after her husband. Before she died, she passed on the duty of acting as Thomas's guardian, which had been bequeathed to her by her late husband, to four relations, Sir John Newton, Charles Bertie, Sir Edward Coke, and John Coke.[2]

Their management was careful and successful and it made the estate richer, stronger, and larger. For an estate, a minority may be a time of recuperation and improvement. A child was cheaper to maintain than an adult landowner, and conscientious guardians were likely to scrutinize the management of an estate with more care and foresight than most landlords responsible only to themselves. As one of the guardians wrote to another: 'In instances of such trusts as we are engaged in, the managemt. should be under Check and controll as much as the nature of the thing would well admit of.'[3] The guardians repaid debts, saved money, and raised the rents. Edward Coke, Thomas's father, had left debts of £22,000, all of which were repaid by the guardians.[4] In 1707-17, £3,626 was saved

[1] In 'Guardians Accounts' (Muniment Room), 'A Comparison of ye Rents of Mr. Cokes Estate as they stood upon the Rental at Michas (1706) . . .', A/B 1708-17.
[2] P.R.O., C. 5/232/8: a Chancery case which served to legalize this transfer.
[3] Holkham Lib. MS. 726, f. 102.
[4] Ibid., MS. 727, ii, f. 449. Edward's personal estate, of about £15,000, contributed a good deal, and a mortgage debt of 28 July 1675 of £5,000 to the duke of Leeds, with interest at 6 per cent, was repaid from the proceeds of a sale of mature timber agreed to by Edward before his death. P.R.O., C. 5/232/8; Holkham Lib. MS. 727, ii, f. 340.

and lent, £555 on mortgages to borrowers in Holkham and £3,070 on government securities.[5]

The estate did not suffer seriously, under the guardians, from spending on the maintenance of relations. There were, however, the substantial properties, worth at least £1,200 a year, in the hands of Lady Anne Walpole, Thomas's grandmother, as jointure.[6] Two marriage portions, of £5,000 each, became due by Edward's will when Thomas's sisters, Cary and Anne, married in 1716. These portions were allowed by the respective husbands to remain unpaid as debts secured on the estates.[7] Thomas had also two younger brothers. They were mainly provided for from the estate of their mother; this estate, worth about £660 a year, went to the second brother, Edward, but was charged with £2,000 for the youngest brother Robert, and £500 each for Cary and Anne. All four younger children were each entitled to annuities from the estate of the young head of the family. The sons got £200 a year for life, the daughters £250 a year until they received their portions. The guardians were able to save money on behalf of the younger children out of these annuities, after paying for their maintenance.[8] In 1720 Thomas Coke bought a commission, evidently as captain, for his brother Robert, for £1,600. A letter survives from 1723 in which one of the former guardians, his grandfather, urged Captain Coke not to apply to him for money, explaining that Thomas had intended that he should live out of the profits of his commission and suggesting that he should suit his expenses to his comings in.[9] Younger children involved no further expenses to the estate during nearly the whole of the rest of the eighteenth century, a great source of financial strength.[10]

On the other hand, the estate under the guardians was burdened with one charge proportionately larger then than it was ever to be again in the whole period of this study, taxation. In the years of the

---

[5] 'Guardians Accounts', 'Money lent on Security'.
[6] H.F.D., 28B and 43. She was a daughter of the duke of Leeds; after the death of her husband, Robert Coke, she had married again.
[7] Holkham Lib. MS. 727, ii, ff. 465-6; H.F.D., 38, 39.
[8] Holkham Lib. MS. 727, ii, ff. 465-6.
[9] Ibid., f. 517
[10] As we shall see, the estate of the younger branch of the Cokes, in Derbyshire, reverted to the elder branch in time to be used to support one after the other of Coke's younger brothers (Edward had them 1727-33, Robert 1733-54) and eventually to be used to support Coke of Norfolk's younger brother. In 1733, too, Robert Coke became Vice-Chamberlain to the Queen.

Estates and Finances under Guardianship, 1707-1718

guardianship £10,292 was paid, or allowed to tenants, for taxes, a figure amounting to 13·3 per cent of the gross income for those years of £84,938. In the years 1708-10 the Land Tax, at its usual war-time level of 4s. in the pound, took away 17·3 per cent of the gross rents due from the farms in Norfolk. The proportion of income taken away by taxation was therefore very large indeed; the burden of taxation on the Coke estates was never again so great until the Napoleonic wars, a century later—and even then it was very slightly smaller. Peace brought relief. Taxes to Michaelmas 1714 were half the figure of taxes for the year ending Michaelmas 1712. Thereafter, as rents increased, and Land Tax assessments remained static, the proportionate impact of the Land Tax declined; the Seven Years War was less burdensome than the wars of 1739-48, and the War of American Independence was proportionately cheaper still for the Coke estate.[11]

In the long run taxes were borne jointly by landlord and tenant. They were, in effect, a charge on the profits of a farm deducted before those profits were divided into landlord's income and farmer's income. But intermittent increases in the rate of taxation would hurt landlords, if, as on the Coke estates in Norfolk, long leases went together with landlord's responsibility for parliamentary taxation, which the tenants paid and then deducted from their rents.[12]

In spite of these outgoings, substantial sums of money remained which could be used to improve the estate. Throughout the seventeenth century the rents had remained fairly stable. On 24 April 1708 the guardians ordered the production of a document showing rents of certain parts of the estate for 1629, 1651, 1656, 1667, 1677, and 1706.[13] For some Norfolk estates of which the acreage is given the rents are shown in Table A.

[11] 'Guardians Accounts', 'An abstract of the Following Accompt . . .', A/B 1707-17. See my note in *English Historical Review*, Apr. 1956.
[12] Intermittent increase would hurt the farmers if, as for instance on the Temple Newsam estates in Yorkshire, leases directed such taxes to be paid by the tenants without deduction from rents (e.g. Leeds City Library, T. N. Altofts and Warmfield with Brockholles, iii. 46. Twenty-one-year lease of 25 Feb. 1707). There was a third possibility: a lease of 1732, granted by Sir Henry Goodricke in Yorkshire, shows that the tenant was to pay taxes unless they rose above 2s. in the pound, when the landlord would pay half (Leeds City Library, Ribston R/120). That would mean sharing the cost of making war.
[13] Holkham Lib. MS. 743, p. 70; H.F.D., 37; J. E. Thorold Rogers, *History of Agriculture and Prices*, vi (Oxford, 1887), 716, printed information from this document.

TABLE A

Township	Acreage	1629	1651	1656	1667	1677	1706
		£	£	£	£	£	£
Billingford	931	275	281	281	281	281	231
Weasenham	961	346	363	359	382	374	374
Fulmodestone	1,041	280	279	279	377	260	300
Godwick and Tittleshall	2,555	640	660	665	660	670	634
Holkham	1,320	302	294	294	291	221	221
	6,809	1,843	1,877	1,878	1,991	1,806	1,760

Rents were shown to be stable during the seventeenth century, but tending to fall towards the end of the century. The guardians succeeded, by contrast, in starting the rise in rents that continued, without a break, until the depression after the Napoleonic wars.

The guardians and their stewards raised rents both by direct increases in rents per acre and by buying new land, making purchases which often increased the value of existing Coke land. Under the guardians, 1707-17, about £12,500 was laid out on purchases.[14] The lands bought, mostly fairly small parcels, adjoined, or were intermingled with, existing Coke properties. For instance, one of the larger purchases was made, about 1709, from Lord Yarmouth, the lordship of the manor, and lands in Sparham, worth £122 a year, for £2,260.[15] The steward advocated the purchase of this estate on sentimental and practical grounds, in order 'to make Mr. Coke's the more intire', and because it was from there 'that my Lord Coke begins his Family in King Edw. the 2d. Raign'.[16] The lands of this estate were 'mingled with Mr. Coke's Ancient Estate wch. by the convenience of the purchase' brought in rent increased by no less than £70 a year, an indirect benefit supplementing the directly purchased rents. There were other similar acquisitions, for example, land in Holkham, 'adjoining to the home grounds' worth £70 a year was bought at 20 years' purchase in 1716, and a house and 43 acres 'Intermix'd with Mr. Cokes' in Castleacre worth £10 a year was bought for £180 in 1717.[17]

[14] 'Guardians Accounts', 'State of the Land purchased by Mr. Coke's Guardians'.
[15] This was 18½ years' purchase—apparently regarded as no bargain price by the steward, who wrote to a guardian in August 1709 recommending that 18 years' purchase be given 'rather than to lett another person in' (Holkham Lib. MS. 727, ii, f. 399). [16] Ibid.
[17] 'Guardians Accounts', 'A comparison of the Rents of Mr. Coke's Estate . . .', and 'A state of the land purchased . . .'.

Estates and Finances under Guardianship, 1707–1718

The Cokes continued to make such acquisitions throughout the eighteenth century. In this way, in parishes where a large owner was concerned to round off his holdings, the ownership of land became more concentrated. If profits were ploughed back and the rents from existing properties were used to pay for new purchases, large estates grew larger; if lands away from the main body of the estate were sold to pay for new lands, the estate did not grow, but became more compact. In either event, the lands in the area of the new purchases could be more efficiently managed; piecemeal enclosures could be constructed and general enclosures more easily arranged; and the landlord could more readily plan farms of the most economic size and the most convenient shape.

In the final accounts of the guardians, there was included a table comparing rents as they were, in 1706, before Thomas's father's death, and as they were, at Michaelmas 1717, just before Thomas came of age. The table showed how much the rental of the estate had been increased by purchase, falling in of leases, or by 'management'.[18] Deducting quit rents, paid out for certain estates, gross rents increased from £4,538 to £5,551 for Norfolk lands alone, and from £5,525 to £6,832 for the whole estate. Those are increases in rents of about 22 per cent. In Norfolk improvement 'by management' accounted for just over half the increase in rents, elsewhere for about one-fifth. On the Norfolk estates 'management' brought about an increase of nearly 12 per cent in rents in the eleven years covered by the table in the guardians' accounts.

Attentive management and rising productivity of land in north-western Norfolk combined to produce the increased rents. These increases were not merely the result of the falling in of very long leases. At the most the farms in Norfolk were let for twenty-one years. Many such leases did fall in under the guardianship. One guardian wrote to another, 'Many leases of the best Farms in Norfolke upon the Rack expired this last Michaelmas; and which are capable of being putt upon a better foot to the Heirs Advantage, both as to the advance of the rent and the putting more repairs upon the Tennant.'[19] The central feature in efficient estate management is to assess accurately the potentialities of the land,

[18] 'Guardians Accounts', 'A comparison of the Rents of Mr. Coke's Estate . . .'.
[19] Holkham Lib. MS. 746, f. 70. 'Upon the rack' signifies land let at an annual rent covering its full value.

to encourage tenants to attain them and to secure a rent correctly reflecting those potentialities. Too low a rent reduces the landlord's income, too high a rent forces farmers to mismanage their farms and harm his land. The guardians closely supervised the application of these principles.

Their supervision was made easier by the introduction of a lucid form of estate account. It appeared in 1707 and was improved in 1711. This was the annual 'Audit Account', the final statement, year by year, of the moneys received and paid out on the estates. Each parish or group of parishes containing Coke properties was accounted for separately. The money due from each individual in each area was first recorded, and, together with the arrears due from the year before, it formed the 'Charge', the sum for which the agent must give account. There followed the 'Discharge' in which all items of expenditure for that area were noted. These items, added to arrears due at the closing of the books, together with the 'neat' or 'net' money (that is, the amount actually passed on by the agent as net income) balanced the 'Charge'. For each estate these same details were given, and, after they had all been dealt with, the information was summarized in a table, the 'Abstract' of the account. For each estate, one after the other, totals were inserted of old arrears, yearly rents, casual profits (forming the 'charge'); rents paid out, taxes, courtkeeping expenses, bailiffs' fees, repairs, improvements, deficiencies, 'neat money', and arrears remaining due (forming the 'discharge'). After the abstract came the 'Account Current', a statement of what the agent had done with the moneys left in his hands—of which the 'neat money' from the estate was the largest part. The annual audit account was kept in this form, with changes in detail, for two centuries.[20]

The minute-book recording the meetings of the guardians, and their decisions, survives at Holkham.[21] They met often, twenty times in 1708, for instance. On matters of general policy, they gave instructions that farmers should live in the house belonging to their farm and that land should not be let to men who had land of their own near to it. The purpose was to prevent tenants giving less care to their landlord's land than to their own. To prevent extravagance,

---

[20] A similar plan of setting out accounts was published in 1715 by George Clerke: *The Landed-Man's Assistant: or, the Steward's Vade Mecum. Containing the newest, most plain and perspicuous Method of keeping the Accompts of Gentlemen's Estates yet extant* (London, 1715), 48.   [21] Holkham Lib. MS. 743.

Estates and Finances under Guardianship, 1707-1718

repairs were to be done under the steward's direction and not at the whim of the tenant, and the stewards were to endeavour to make tenants pay for repairs, except for necessary timber, which was to be laid out by the steward. The stewards were to prevent tenants subletting parts of their farms 'to the prejudice of the Estate'; subtenants would normally pay a higher proportionate rent than the principal tenant, and either this rent should have gone to the landlord or it must have been raised by overworking the land sublet. The guardians decreed that before any increase or reduction in rent took place a survey and valuation of the farm should be submitted to the guardians, to end the practice of using former rents as a guide. Several painstaking estate maps date from this time. On the same occasion the guardians instructed the stewards to produce a report on the condition of 'the Tenants and their Farmes ... for the better distinguishing between the good and bad Husbands in order to the encouraging the one and timely admonishing the other'.[22]

Money might be the 'encouragement'. When the chief steward, Humphrey Smith, wrote to give information on the condition of the farms at Sparham, when their purchase was proposed, he remarked of a farmer there, 'Murrell is a good husband and Improves his land by Marling at his owne charge. Doe not know but at first my Lord might give some incouragemt.'[23] The encouragement would be for the landlord to bear a part or the whole of the expenses of marling—a process likely to benefit the land for some decades.[24] The audit books show deductions, allowed from rents, for the costs of marling on several farms, evidently as a result of contracts made between landlord and tenants. During the five years ending at Michaelmas 1715 John Carr of Massingham marled by agreement 240 acres, and was allowed 8s. an acre in return, a total of £96. (Carr held his farm at a rent of £135 per annum.)[25] In this and other ways the guardians invested liberally in the estates under their care; in the years 1707 to 1717 at least 15 per cent of gross rents was returned to the estate in repairs and improvements.[26] Such

---

[22] Ibid. pp. 10, 11, 79, 64, 76, 77.
[23] Ibid., MS. 727, ii, f. 399.
[24] The nature of marling is discussed below, p. 8.
[25] A/B 1707, 1711-15, Massingham. There are other similar entries in this period, e.g. A/B 1713, Fulmodestone, F. Etteridge.
[26] 'Guardians Accounts', 'An Abstract of the Following Accompt'. Very little of this was laid out on the Coke estates outside East Anglia.

investment sometimes gave a direct and immediate return; from Michaelmas 1711 a tenant in Weasenham paid an extra £10 a year 'for the interest of £200 laid out in building a barne and which is to remain an improved rent'.[27]

Careful management, willingness to invest money, and the use by tenants of new farming techniques, explain the rise in rents that began when the guardians were responsible for the estates. Under the guardianship, everything that was needed to make possible the progress that took place on the Coke estate in north-western Norfolk in the eighteenth and early nineteenth centuries was at hand: marling, turnips, clover, and artificial grasses. Together with wheat and barley, it was turnips, sown grasses, and clover that made up the crop rotations, which culminated in the famous four-course, and which brought Norfolk farming to great heights of prosperity and renown. Under the guardianship their use had begun, and the results, in rising rents, had started to appear.

Marling, or claying, was a process involving digging up certain useful types of subsoil and mingling it with the topsoil. The effect, especially on poor light soil, was highly beneficial, one application fertilized and strengthened the topsoil for a period of some years: 'its benefit, though not to the effect it produces at first, is felt for thirty years, when a second marling, of about half the original quantity, may . . . be used.'[28]

Turnips were the essential factor in the advance of Norfolk farming. They were essential for feeding cattle and sheep in the winter. They brought direct benefit, therefore, in increasing the amount of livestock it was possible to maintain, with the consequent indirect benefit of providing more animal manure. On light soils turnips are easily cultivated—and most of the Coke lands in Norfolk were composed of light, and often very light, sandy soil. On such soil, too, sheep could usefully be fed with turnips on the land on which the turnips had been grown; the effect was to manure the soil and also to improve its texture by the trampling of the animals. Thus turnips had a valuable effect on a succeeding cereal crop. If the turnip fields were adequately hoed, weeds could be controlled as effectively as by the use of unprofitable fallows.

Turnips were being grown in central Norfolk as early as 1650-

[27] A/B 1711-17, Weasenham, John Tubbing.
[28] Nathaniel Kent, *General View of the Agriculture of the County of Norfolk* (London, 1796), 23.

Estates and Finances under Guardianship, 1707-1718   9

60,[29] although there is no sign of turnips at Holkham in 1641-50.[30] It remains difficult to form any clear notion of how far the use of turnips had spread by any particular time. Significantly, Worlidge wrote about turnips in his book, published in 1668:

> Although this be a Plant usually nourisht in Gardens, and be properly a Garden Plant, yet is to the very great Advantage of the Husbandman sown in his Fields in several forreign places, and also in some parts of England, not only for Culinary uses . . . but also for Food for Cattel. . . . They delight in a warm, mellow, and light land, rather sandy than otherwise . . . It is a very great neglect and deficiency in our English Husbandry that this particular Piece is no more prosecuted.[31]

Probably north-west Norfolk was one of the parts of England he referred to. Defoe declared that turnips were first used as fodder in east Suffolk 'from where the practice is spread over much of the east and south parts of England', that is, by the 1720s.[32]

We can safely assume that the use of turnips had become established practice in north-west Norfolk by the early years of the eighteenth century. Dr. Plumb has written that turnips were being grown regularly at Houghton 'in very considerable quantities' by 1673, and has shown that turnips were being grown on various parts of the estates of Robert Walpole in north-west Norfolk, apparently as a matter of course, in 1701.[33] The same is true of the Townshend estate, which in many places adjoined Coke properties, in 1706.[34] Turnips were growing at North Elmham, near to, if not on, Coke land, in 1707, and similarly at Massingham in 1718.[35]

Unfortunately direct evidence of what Coke tenants were growing is rare. But turnips were being grown on the Honclecronkdale farm

[29] J. Spratt, 'Agrarian Conditions in Norfolk and Suffolk, 1600-50', M.A. Thesis (London University, 1935), 203. He concluded that 'root crops were still in the experimental stage in 1650'.
[30] Tithe payments used by Spratt, p. 195.
[31] J. W. W[orlidge], *Systema Agriculturae* (London, 1668), 46.
[32] D. Defoe, *Tour through England and Wales* (1st edn. 1724-6), Everyman's Library, 2 vols. (1928), i. 58. E. Kerridge in *Economic History Review*, 2nd Ser. viii (1956), 390-2, has produced evidence to support Defoe; Kerridge attributed the introduction of turnip husbandry to the years 1646-56.
[33] J. H. Plumb (ed.), 'The Walpoles: Father and Son', *Studies in Social History, A Tribute to G. M. Trevelyan* (London, 1955), 184; J. H. Plumb, 'Sir Robert Walpole and Norfolk Husbandry', *Economic History Review*, 2nd Ser. v, no. 1 (1952).
[34] H. W. Saunders, 'Estate Management at Raynham 1661-86 and 1706', *Norfolk Archaeology*, xix (Norwich, 1917), 63-6.
[35] Norwich Castle Archives, Case 33, Shelf E/7 and F/15. Inventories exhibited at Norwich Archdeaconry Court, 1707, 1718.

at Holkham in 1709, when it was in hand for a year, and when it was let the tenant agreed to leave, at the end of his tenure, 15 acres of 'Somerland plowed and Harrowed in the third Earth for Turnapps, as the land shall come in course'.[36] At Fulmodestone, in 1709, £12. 17s. 0d. was paid to an outgoing tenant for 18 acres of 'Summer Land Plough'd in the 3d. Earth for Turnipps', and the same amount was to be left by the next tenant.[37]

A difficulty in dating the spread of turnip husbandry comes from the fact that turnips appear at first to have been grown as fodder for stalled cattle. The contribution of the turnip, however, to the development of Norfolk husbandry was in its use as part of an arable rotation, when it was eaten by sheep or cattle on the land on which it grew. The discovery of turnip cultivation in a particular place at a particular date does not necessarily imply that turnip husbandry was fully developed there at that time. The evidence from the Coke estates, however, makes it plausible to suggest that turnips had been grown in north-western Norfolk for some decades but that their use in a developed turnip husbandry, though a firmly established innovation, was still in its early stages in the first years of the eighteenth century.

Clover enriches the soil with nitrogen and thus provides plant food for subsequent crops. Clover and sown grasses made it possible to use arable land for maintaining livestock. Their use enabled an increase in the number of livestock on a farm without reducing arable acreage or an increase in the arable acreage without diminishing the quantity of livestock. Thus their use facilitated the manuring of the soil by making it easier to maintain livestock, and also directly improved the quality of the soil for cereal crops.

By the end of the seventeenth century clover was in use on both the Walpole and the Townshend estates[38] and there is evidence to show that clover was being grown on the Coke estates by the time of the guardianship. A small cultivator left some clover in his fields at Weasenham when he died in 1708 and another did so at Bylaugh in 1712.[39] At Michaelmas 1709 there were 24 acres sown with clover on the Honclecronkdale farm at Holkham.[40] A tenant taking

[36] A/B 1709; Holkham Deeds, 1007.
[37] A/B 1709.
[38] Plumb, *Economic History Review*, 2nd Ser. v, no. 1; Saunders, op. cit., 63, 65-6.
[39] Norwich Castle Archives, Case 22, Shelf F/8, 88 (John Climence) and Shelf F/11, 38 (William Watts).
[40] A/B 1709.

a farm in Castleacre in 1714 was forbidden to take above four crops before laying down with clover and nonsuch.[41]

The growing of wheat was not a cause of agricultural improvement in this area, but it has often been regarded as a symptom of it. There is no doubt that the acreage of wheat increased during the eighteenth century. But, on the other hand, wheat was a familiar crop in north-western Norfolk at the time of the guardianship, both alone and in the form of 'mislyn' or 'mixtlin'—almost every conceivable spelling can be found—that is, wheat mixed with rye. At Holkham Robert Franklin had 14 acres of wheat, compared with only 2 of rye, ripening at the time of his death in August 1709.[42] On the Honclecronkdale farm at Holkham 7 acres were under 'meslin' in 1709 and 20 acres were prepared for wheat.[43]

The special suitability of the soil and climate of north-western Norfolk for the growing of barley was already fully understood. Records of crops grown or harvested in the area invariably show a strong preponderance of barley: the Honclecronkdale farm in 1709 carried 67 acres of barley, 33 acres of peas, 15 acres of vetches, and only 7 acres of 'meslin'.[44] A deceased husbandman left behind him in Massingham in September 1718, 28 acres of barley, 12 acres of oats, 11 acres of 'rye and mixtling', 11 acres of peas, $5\frac{1}{2}$ acres of turnips, 5 acres of vetches, and $1\frac{1}{2}$ acres of wheat.[45] These examples are characteristic. Both with wheat and barley one can say that the essential change in eighteenth-century Norfolk was not their appearance in areas where none had been seen, but the extension of their growth within areas where they were already familiar.

The work of the guardians established a trend towards increasing productivity on the farms and efficiency in management which, together with their discharge of debt and husbandry of pecuniary resources, enabled the estate to survive the great strains laid on it by their ward, once he had come of age. The next two sections describe those strains—the financial disaster of 1720 and the building of Holkham Hall.

[41] Holkham Lib. MS. 743, pp. 156-7.
[42] Bishop of Norwich's Archives, Box 121, no. 266.
[43] A/B 1709.
[44] Ibid.
[45] Norwich Castle Archives, Case 33, Shelf F/13, no. 28 (Thomas Burrage).

2

Thomas Coke and the South Sea Bubble

> When Moses with his Army the Red Sea cros'd over,
> Proud Pharaoh and his Troops no danger did discover,
> But rashly pursued, till by Waves they'r Surrounded,
> And he with his Host in an instant was Drowned;
> Thus thousands of late have past the South-Seas,
> As safe as in Waters not up to their Knees,
> Whilst those that came after without Wit or Fear
> Like Pharaoh's Great Host are now nick'd in the Rear.[1]

ON 2 July 1718, shortly after he came of age, Thomas Coke married Margaret Tufton, third surviving daughter of Thomas, earl of Thanet. Her portion was fixed at £15,000.[2] Of this money, £10,000 was used for providing portions of £5,000 for each of Thomas's two sisters, both of whom had married in 1716, Cary to Sir Marmaduke Wyvill and Anne to Philip Roberts. Cary's portion had been advanced in 1716, at 5 per cent, by Sir John Newton, her maternal grandfather and one of the guardians. After the marriage Lord Thanet took over the debt to Newton.[3] The history of the other £5,000 portion is obscure. Thanet started to pay the interest on this capital sum—Roberts, the husband, agreed to leave the capital and draw only the interest—but by 1725 Coke was paying the interest once more, still at 5 per cent. Probably Thanet had paid over another £5,000 in addition to the £5,000 he paid to Coke's cashier on the day of the marriage. The major financial benefit Thomas Coke and his successors eventually derived from the marriage will be discussed later—that was the lease of a lighthouse. The immediate result of his marriage was that the estate was spared much of the burden of providing for Thomas's sisters. Thomas

---

[1] *Weekly Journal or British Gazetteer*, p. 1707 (Sat., 10 Sept. 1720).
[2] Holkham Lib. MS. 742 contains a copy of the terms of the marriage settlement.
[3] Ibid., MS. 741 shows Thanet paying the interest on it.

Thomas Coke and the South Sea Bubble 13

Coke's debts immediately after his marriage cannot have been much more than £7,000 or so.[4]

Even before the disaster of 1720, Coke's debts had mounted as a result of a spirited burst of extravagance in his first two years in control of his estate, which led to his overspending by £15,000 or so. It is difficult to know how he did it. Certainly £2,210 went to pay the bills of 'Mr. Lamere, the silversmith' and £4,126 was spent on jewellery. Some of the spending was more productive: £1,000 was paid to Mr. Kindersley in part of £2,000 for draining a section of the Holkham marshes.[5] The money was lent by Peniston Lamb, Thomas Coke's scrivener, cashier, and banker. On 1 April 1720 Lamb was given a mortgage for £15,000 at 5 per cent interest.[6] It was with his finances thus weakened that Coke was caught up in the feverish outburst of speculation of 1720.

There is no need to describe here the South Sea Bubble episode or to explain the causes of the speculative fever of 1720.[7] Exact knowledge of gains made, or losses sustained, by individuals through transactions in South Sea stock is rare. It is fortunately possible to trace Coke's involvement in the South Sea mania, and to work out his precise losses. He provides one example of the effect the events of 1720 could have on a landowner's position.[8]

Thomas Coke first acquired South Sea stock by taking up some of the new issues. He subscribed for £1,000 stock in the first public subscription of April 1720 for new stock offered at 300. Peniston Lamb recorded the payment made by his firm to Coke on 7 June of £600 'to finish his purchase of 1000£ subscription in the South Sea'. On 13 June £300 more was paid and on 15 August another £300, calls of 10 per cent on subscribers. No other payment was made and Coke therefore paid £1,200 on account of the first subscription.

It was probably in June that the transactions were begun which

[4] £7,000 was due to Mr. Roberts: £500 left to Anne, by her father and £1,500 saved by the guardians from her maintenance money in addition to her £5,000 portion.
[5] Holkham Lib. MS. 741, ff. 3-7.   [6] Ibid., f. 8ᵛ.
[7] The best account is P. G. M. Dickson, *The Financial Revolution in England* (London, 1967), 90-198. See also W. R. Scott, *The Constitution and Finance of English, Scottish and Irish Joint-Stock Companies to 1720*, iii (Cambridge, 1911), 288-360.
[8] Holkham Lib. MS. 741 includes an account of the moneys paid and received on Coke's behalf by Mr. Thomas Gibson and partners in 1718, and by Peniston Lamb Esq. in 1718-29, together with a statement of Coke's dealings in South Sea stock. The Lambs, until the latter part of the eighteenth century, acted as solicitors and conveyancers to the Cokes and in the earlier part of the century they were also their cashiers. Peniston Lamb's nephew was Sir Matthew Lamb, whose son and grandson were Peniston, the 1st Lord Melbourne and William, the 2nd Lord Melbourne, who became Prime Minister.

were recorded by Lamb in these words: 'To Mr. Darcy £3,200 which with 2000£ the Difference of £1000 South Sea Stock between 680 & 880 & also with £1600 the Difference of another £1000 stock between 680 & 840 making together £6800 for £1000 South Sea stock bought of Mr. Darcy at £680 transferd to Mr. Snow in Trust for Mr. Coke & paid Mr. Snow £10 commission. In all 3210.0.' The stock, if bought at the market price, was bought about 1 June, and sold some time between 4 and 24 August; Lamb's entry is dated 19 August. Apparently Coke's agents agreed to buy £3,000 stock at 680 in June, of which £2,000 were sold later on in order to make a profit to help to pay for the rest of the stock—£1,000 of it— that Coke went on holding. Probably these were transactions on margin and the £3,210 was almost certainly the only cash paid. The £1,000 stock Coke kept cost him £6,800; £3,600 was met by taking profits on stock bought and the £3,200 made the amount up. Evidently Coke was compelled to pay because he could not get credit beyond 19 August, the date on which he paid. As a result Coke, in effect, secured £1,000 stock for £3,200.

Another purchase made in June was of £500 stock in the third public subscription. This stock was bought in the name of John Coke. Stock was offered by the company at 1,000, 10 per cent of the purchase money to be paid at once, the rest in nine instalments at half-yearly intervals from the summer of 1721.[9] Though the amount shown as paid for this stock is £550, only £500 can have been paid for the stock itself. A second £500 seems to have been invested in the third subscription as the first payment for another £500 stock at 1,000; there is no direct evidence for its purchase, but its existence, and, in due course, its sale, are recorded.

Coke bought no stock in July. But he bought a lot of stock in August, not only South Sea stock but stock in other companies as well. Precise information about these latter dealings does not exist, but there are some details. On 3 August he paid £380 for 5 shares in the 'Gold & Silver'.[10] On 11 August £1,000 was paid for £1,000 stock in the 'London Insurance', stock bought at 100, representing a large premium on the then paid-up capital of £5 per share of £100.[11] On 20 September a call of 5 per cent was paid on this stock, another £50. On 19 August another £110 was paid for subscription to

[9] Scott, iii. 320.
[10] Probably this was the 'Company for extracting gold and silver from lead and other sorts of ores' for which subscriptions opened on 2 Aug. (ibid. 457). [11] Ibid. 411.

Thomas Coke and the South Sea Bubble 15

London Insurance. What happened to this stock cannot be found. A definite loss in these speculations in stocks other than the South Sea appears in November when £2,257 10s. 0d. was paid to Mr. Snow for 'the Difference of £3000 Ram.[12] bought at £125 and sold at £50'. There was paid to Snow at the same time £135 'for the use thereof for 2 months from 25 Septr. to this day at 3£ p.cent p.mensem'.

Perhaps this loss was partly balanced by a transaction of which only one part appears in Lamb's accounts, the receipt, on 3 August 1720, of £3,280 for £4,000 London Insurance sold at 82. It cannot be said whether gains or losses came from these dealings, which were certainly less important than the dealings in South Sea stock. It is more likely that they led to losses than to gains; any gains must have been small.

Coke's main interest continued to be in South Sea stock, and on 19 August 1720, he made his most disastrous purchases. On that day, he bought £1,500 stock from Dr. Hugh Chamberlen[13] at 970, and £1,500 stock from Mr. Snow,[14] also at 970. These purchases, made after the price of South Sea stock had begun to fall, cost £29,100.

He bought still more stock in September, after the collapse had begun. In that month, Coke bought £2,000 South Sea stock at 680 from Daniel, Lord Finch,[15] and £1,500 stock, also at 680, from Sackville Tufton.[16] The cost was £23,800.

By the end of September, then, Coke had laid out at least the following total in purchases of stock:

	Money laid out £
In the first subscription in South Sea stock	1,200
For Darcy's stock, a balance of	3,200
In the third money subscription in South Sea stock (£500 not quite certain)	1,000
For 'Gold & Silver' shares	380
For London Insurance stock	1,050
The August purchases of South Sea stock	29,100
The September purchases of South Sea stock	23,800
	£59,730

For South Sea stock alone the total was £58,300.

[12] i.e. London Insurance.
[13] The fashionable London obstetrician. He inherited the family secret of the use of midwifery forceps.
[14] Snow was a goldsmith-banker. P. G. M. Dickson, op. cit., refers to other of his dealings.
[15] Eldest son of the 2nd earl of Nottingham.
[16] Later 7th earl of Thanet, Coke's wife's cousin.

Why were most of Coke's purchases completed in August and September, after the fall in South Sea stock had begun, rather than in May, June, or July? Possibly the purchases of August and September were the result of 'future' contracts; possibly, that is, Coke had agreed at an earlier date to buy an agreed amount of stock at a settled price on a particular day. This would explain why South Sea stock was bought on 19 August at a price (970) far above the market price of that day (840-50, according to Scott). The September purchases, on the other hand, at 680, were not above the market price if they were made before 8 September. The times when Thomas Coke bought stock could be explained by two possible reasons; either his eagerness to buy and hold South Sea stock fluctuated, or there were changes in the ease with which he could buy and hold stock. But Coke seems to have been very eager indeed to buy and hold stock throughout the critical months, May to September, of 1720. He subscribed in April, and again, on onerous terms, in June. Probably in June, again, he bought stock from Darcy.

The dealings with Darcy are significant. He bought from him more stock (£20,400 worth—£3,000 stock at 680) than the £1,000 he was eventually able to hold and he had to resell the rest of it by 19 August. Evidently when he sold this stock, he was not selling willingly, for he kept some of Darcy's stock and bought and kept other stock about the same time at which he sold stock to pay Darcy. It is true that Coke made a profit on the part of Darcy's stock he sold; but if he had been looking for immediate profit, as distinct from greater profits in an illusory future, he might reasonably have sold all Darcy's stock. It is true that he could only have sold the stock he acquired on 19 August at 970 at a loss; but if he had, in fact, sold Darcy's stock voluntarily at the prices he did (840 and 880) in order to make profits, this would mean that he did not think it likely that the price of stock would rise again and make greater profits possible later. If he *had* made this pessimistic assumption, unless his actions were wholly illogical, he would also have sold the stock bought at 970 in order to avoid greater losses on it in the future. Apparently, then, Coke had been, and remained, anxious to hold stock at the time he resold part of the stock he bought from Darcy; the sale of part of Darcy's stock cannot have been desired by Coke, and we may reasonably ask why he was able to buy and hold more stock in August and September than he had been earlier.

Coke's dealings with Darcy, and the manner in which he bought and held more stock in August and September, strongly suggests that relative ease of buying and holding stock on credit explains the timing of Coke's activities in the stock market in 1720. He sold, it seems reasonable to argue, some of the stock he bought from Darcy, because he could not continue to hold it without getting credit; instead, Darcy wished to be paid. All the stock Coke bought from Chamberlen and Snow, and all the stock he bought later from Finch and Tufton was bought and held on credit. Probably whether credit was available or not explains the timing of Coke's purchases, and it may well be that Coke's losses in the South Sea disaster were limited only by the amount he could borrow in the fatal summer of 1720. It is not surprising that there were sellers of South Sea stock in the market in August and September, when the danger signals were in sight, ready to give easy credit to a rich landowner, who might be overspending his income, but whose ultimate capital resources were very large. It is possible that this made the South Sea crisis into a peculiarly dangerous trap for landowners, as more sophisticated speculators moved to convert their gains into first class long-term loans, secured on landed property.

Borrowing liquid funds was very expensive in 1720. In June £120 was paid on Coke's behalf to Mr. Snow 'for the use of £3,000 for a month at 4 p.cent' (that is, 4 per cent *per month*) and in November £135 was paid for £2,257. 10s. 0d. for two months at 3 per cent per month. Archibald Hutcheson, in a preface to a pamphlet of April 1720 discussing the inflation in the price of South Sea stock, hoped that there would soon end 'the Borrowing of Money, at the rate of 10l. per Cent per Mensem; and even at 20s per Cent per Diem'.[17] Such shortage of credit supports the suggestion that Coke may have bought and held stock when he did because it was only then that he could get credit to do it. On the other hand, Coke borrowed £6,000 at 5 per cent from Henry Portman on a mortgage dated 7 June 1720. But such a mortgage would have been agreed on and contracted for some time before the necessary legal formalities could be completed.

The prices of Coke's purchases were reduced by his creditors when his adventures in 1720 came to be paid for. For Chamberlen's

---

[17] Archibald Hutcheson, 'A Collection of Calculations and Remarks relating to the South Sea Scheme & Stock', preface to *Some Seasonable Considerations for those, who are Desirous . . . to become Proprietors of South Sea stock . . .* (London, 1720), 25.

stock, we find that the 'Doctor accepts of the same at the price £700' instead of 970, and Snow reduced his price from 970 to 800. Finch and Tufton agreed to accept 600 instead of 680 for the price of the stock they had sold. So in working out Coke's losses, something must be deducted from the £58,300 given above as his total spending on South Sea stock. That was the figure he would have paid if full payment had been exacted. As it was he paid:

	£
The first subscription	1,200
Purchase from Darcy, balance of	3,200
Third subscription	1,000 (of which £500 is doubtful)
Finally paid for Chamberlen's stock	10,500
Finally paid for Snow's stock	12,000
Finally paid for Finch's stock	12,000
Finally paid for Tufton's stock	9,000
	£48,900

The total face-value of stock Coke acquired, and the prices at which he sold it must be found in order to say what he lost. The £1,200 paid to the company on account of the first subscription of 1720, after changes in the terms of subscription and the giving up of any further calls on subscribers, was turned into stock at 300. Thus Coke secured £400 stock, his cheapest acquisition; subscription was much less ruinous than purchase at the worst times. The 1720 midsummer dividend, of 10 per cent in stock, added another £40 stock. Then, in consequence of the distribution, under parliamentary sanction, of £33. 6s. 8d. for each £100 stock held, this grew by another £146. 13s. 4d. to a total of £586. 13s. 4d.[18] The £1,000 invested in the third money subscription of 1720 was eventually made into stock at 400, making £250 stock. The midsummer dividend of 1720 made this £275. The first distribution of the summer

---

[18] D. Macpherson, *Annals of Commerce*, iii (London, 1805), 109-11, gives a useful and concise statement of the settlement made by parliament and the company after the slump. He is wrong in suggesting that the first money subscription was, like the other subscriptions, at once turned into stock at 400. He is wrong also in saying that the first money subscribers shared in distributions of stock to the same extent as the other three money subscriptions. There were two such distributions in 1721, the first paid by direct parliamentary enactment (7 Geo. I, s. 2), the second by decision of the General Court of the Company, acting under the general direction of the Act. Both these distributions added one-third to the stock of the proprietors in question. The first money subscribers did not share in the first distribution under the Act to restore Public Credit.

Thomas Coke and the South Sea Bubble

of 1721, of £33. 6s. 8d. per cent of stock, brought it to £366. 13s. 4d. and the second distribution made it £488. 17s. 8d. The stock bought from Darcy, Chamberlen, and Snow was lumped together. When the 10 per cent stock dividend of midsummer of 1720 was added to the total face value of this stock of £4,000, it became £4,400. (Stock sold in the first part of August 1720 was sold inclusive of that dividend.) This was handed over to Snow to be sold by him at any time he should think fit before 1 August 1721. To this stock was added the second 'parliamentary' distribution of £33. 6s. 8d. stock per cent ('old' South Sea stock, created before 1720, did not qualify for the first distribution) and that made £5,866. 13s. 4d. face value in stock. The £2,000 stock bought from Lord Finch was evidently bought too late for it to rank for the midsummer dividend of 1720, and £1,000 of it was sold before any bonus stock was added. The other £1,000, however, grew to £1,333. 6s. 8d. The stock bought from Sackville Tufton similarly did not qualify for the dividend of 1720, but, supplemented by one-third, it became £2,000 in face value.

At one time or another in 1720 and 1721, therefore, Coke owned a total of £11,275. 11s. 0d. of stock, of which £1,000 was sold in 1720. This was sold by Lord Finch, with Coke's consent, at 170, the best price of any of the sales. It brought in £1,700. The whole of the remaining £10,275. 11s. 0d., except £488. 17s. 8d., was sold in October 1721. The prices were between 89 and 92½ and the sum realized was £8,827. 3s. 4d. The remaining £488. 17s. 8d. stock was not sold until May 1722, when, at 90¾, it produced £443. 12s. 0d. The sales of all the stock acquired in 1720 brought in, therefore, a total of £10,971. 5s. 4d.[19] This stock had cost £48,900. Thomas Coke, then, lost £37,928. 14s. 8d. by his dealings in South Sea stock.[20]

The dividends paid by the company must be set off against this loss of capital. All Coke's stock, except the £1,000 sold before the rest, was entitled to the 5 per cent Christmas dividend of 1720 and the 4 per cent midsummer dividend of 1721. The total is about £685. The 3 per cent dividend of Christmas 1721 was paid on the stock that was kept until May 1722 including dividend on the bonuses added to this stock in 1721. The dividend payment was

---

[19] Of this it must be recalled, £221. 16s. 0d. is uncertain: the product of the uncertain investment of £500 in the third subscription.

[20] Or, if the doubtful half of the third subscription is removed, £37,650. 10s. 8d.

£14. 13s. 4d. Interest payments on the debts incurred to pay for stock were certainly far higher than the dividends received from the stock.

Almost the whole of the money received when stock was sold was used to help to pay for its original purchase. Lord Finch took the £1,700 for the stock he sold and was given another £1,700 which Coke borrowed from Edmund Waller,[21] who was repaid later out of the proceeds of sales. Snow was given £5,600, which reduced the debt to him to £6,400. There remained due to Finch £8,600, to Sackville Tufton £9,000, and to Chamberlen £10,500. Snow took a bond; Finch was given a bond and mortgage dated 13 January 1721; Tufton got a bond for £3,000 to be paid on 1 March 1721 and his remaining £6,000 was secured by bond and mortgage of 13 January 1721; Chamberlen's money was secured to him by mortgage and judgement from 21 December 1720. The mortgage debt due to Lamb rose to £24,000.

The events of 1720 help to explain why the Cokes never plunged again into the stock market, never again bought securities to more than a negligible amount during the whole period of this study.

[21] He was John Aislabie's son-in-law. See Dickson, op. cit., 129, n. 1.

3
General Finance, 1722-1759: the Building of Holkham Hall and the Exploitation of the Lighthouse

THOMAS COKE'S financial position after the South Sea crash was uncomfortable, and seemed to leave no room for major items of new spending. In the year ending at Lady Day 1723, Coke overspent by £992. 13s. 8d. Even this unsatisfactory result was arrived at only because of the work of Coke himself at the gaming tables. He won £2,592. 8s. 6d. at play that year and lost only £146. 10s. 0d., so that his clear gain was £2,445. 18s. 6d. This sum, together with £7,802. 10s. 11d. net revenue from the estate, made up nearly the whole total receipts of £10,501. 9s. 5d.[1] The largest single item in the total payments of £11,494. 3s. 0d. was, as one would expect, the amount paid in interest on debt: no less than £2,564. 19s. 9d. A total of £7,904. 0s. 3d. went on expenses of living. Of this, £1,172. 11s. 6d. went on housekeeping, plus £626. 18s. 8d. on 'pantry and cellars'. Stables and hounds cost no less than £1,482. 12s. 9d. A figure of £1,229. 6s. 6d. was used up as 'Pockett' money and £740. 5s. 10d. went on books and furniture. Only £417. 9s. 4d. went on servants' wages and liveries (some wages, including the estate steward's salary, were paid before the net estate revenue was arrived at) and only £157. 14s. 6d. was spent on clothes. From the accounts it appears that the balance was improved by the simple expedient of delaying the payment of tradesmen's bills: £873. 2s. 1d. due to them was left unpaid.[2] There is no reason to

[1] The money won came mainly from Lord Finch—who perhaps could afford the £1,050 he lost, after his dealings in South Sea stock—and John Harvey who lost £630. Other losers were Sir Humphrey Howarth, Lord Hillsborough, Lord Sunderland, Lord Strafford, Lord Millsington, and Lord Ashburnham. 'Mr. Coke's own Acct. of moneys that have pass'd thro his hands', p. 16 in Ledger book of 'Mr. Coke's Accounts for one whole year ending Lady Day 1723', in Holkham Library.

[2] Ibid. containing 'General Abstract of Expences for 1722 and by whom paid' and 'Account of Profit and Losse; or of all payments and Debts to Expences, and Losses as also of all Receipts, and Creditts to profit For one whole year ending, as to the Land Estate at Michas 1722 in other respects to Lady Day 1723' in Holkham Library.

suppose that Coke spent less, as time went on, on everyday things. Housekeeping at Holkham cost about £2,400 a year at the end of the 1730s.[3] This included payments made for food and drink and for household equipment, wages, fuel, laundry, medicines, and doctors' fees. The stables were costing £630 a year on average, clothes for Lord Lovell,[4] as he then was, over £200 a year, his pocket money about £400 a year. Furniture, presumably in addition to the amounts included in the building accounts, cost about £375 a year.

One point deserves emphasis: the proceeds of the estate appeared in these accounts at the stage of net proceeds, that is, after deduction of taxes and payments for repairs and improvements. The proceeds credited were the proceeds of an efficiently managed estate from which everything needed to keep the estate in good order had already been taken. The fact is expressive of a state of mind. The estate was the basis of all the Coke life and its efficient management was axiomatic. No financial stringency ever led to skimping on the estate. Spending on it, including the buying of small properties, was accounted for as a prior charge, coming before housekeeping or personal needs. Of course, an accounting system has no more than psychological force and certainly could not prevent a Coke landlord from economizing on estate management—but none of them ever did so. Financial difficulty might cause sales of land—though not from the central block of land in north-western Norfolk—but it was never allowed to affect the management of the estate.

Obviously one would expect a rise in income before a beginning was made on the building of the new house at Holkham. The amount left available to Coke from his estate revenue had already been increased by the death of his grandmother, Lady Anne Walpole. Her death in 1722 meant that it was no longer necessary to pay her £1,200 a year—the amount for which her jointure lands had been leased in 1718.[5] Rents rose: in 1720 gross rents (from the whole estate) were £9,101. 17s. 10d.[6] In 1730 they were £11,565. 4s. 4d. In consequence net money rose correspondingly, though its amount fluctuated from year to year: casual profits

---

[3] Holkham Lib. MS. 731.
[4] Coke became Baron Lovell in 1728 and earl of Leicester in 1744.
[5] H.F.D., 43.
[6] Holkham Lib. MS. 726, f. 100.

moved up or down as did arrears and spending on repairs, improvements, and taxes. The average net income from the estate in 1728-32 was £8,875 per annum.[7]

Soon, too, non-agricultural sources began to supplement the income from the estate. Coke began to derive revenue from the most important element in his wife's dowry, the Dungeness lighthouse, in 1730.[8] In 1735 £1,412. 8s. od. clear of all deductions came from it.[9] In 1733 Coke became joint Postmaster-General at a salary of £1,000 per annum.[10] He continued to receive his salary from the Post Office until his death.[11] By 1733, however, building at Holkham had already begun; it is likely that the increased income from the estate and the money from the lighthouse tolls were more important in causing Coke to make up his mind to start building his new house than the extra profits from politics. Supplements to estate income were certainly necessary, for there are signs of financial stress throughout the 1720s; Coke was even compelled to use his plate as security, first to Gibson for a loan of £1,200 in 1726, and later to Edmund Waller to secure £1,600 in 1727.[12] Borrowings followed one another at a hectic rate. Repayments did not keep pace. Though, by then, Coke had paid off £12,350 of his debts of 1721 and had secured reductions of interest on other debts from 5 to 4½ per cent, Lamb paid over £3,700 in interest payments on his behalf in the year May 1729 to 1730.[13]

The building of the great house at Holkham was certainly not undertaken as a consequence of overwhelming prosperity. Coke was very eager to build. In the last six years of his minority, from 1712 to 1718, he had carried out a prolonged and extensive European tour.[14] At that time the future designer of Holkham Hall, William Kent, was studying in Italy. Thomas Coke met him in Rome in 1714, when Kent was 29 and Coke 17. Coke and Kent

[7] A/B 1728-32.
[8] Holkham Lib. MS. 741, f. 46ᵛ. His father-in-law, Lord Thanet, died in July 1729.
[9] Some lighthouse accounts are in the game larder.
[10] B.M. Add. MSS. 36130, f. 298. K. Ellis, *The Post Office in the Eighteenth Century* (London, 1958), 14, shows that, at least by the 1780s, allowances and perquisites brought this figure up to about £2,900.
[11] For a few months, in 1745 and 1758-9, he was sole Postmaster-General. G. E. Cokayne, *Complete Peerage*, ed. V. Gibbs *et al.* (London, 1910-40).
[12] Holkham Lib. MS. 741, ff. 32ᵛ and 37ᵛ.
[13] Ibid., esp. ff. 47ᵛ to 49ᵛ.
[14] C. W. James, *Chief Justice Coke, His Family and Descendants at Holkham* (London, 1929), 179-208.

went to Venice in August 1714 and later visited Padua together. In 1716 Kent met Coke in Sicily and they spent the summer in Rome. They went together to Florence in September. Coke's guardians contributed to Kent's expenses and employed him to buy pictures, marbles, and other works of art in Rome. Coke studied and developed designs for his house in consultation with Kent and Burlington for sixteen years after his return from Italy.[15]

A total of about £92,000 was spent on building the new house and on the creation of park, gardens, lake, and woods. It is difficult to be sure how much should be added for money spent on furnishing the new house. Equally it is practically impossible to say how much of the £92,000 should be regarded as the expenses of upkeep and not as spent on new construction or planting. However, there was not much at Holkham to keep up—the house and park at Holkham were both entirely created in the years in which this money was laid out—there was little or nothing to be extended or modified; everything had to be begun.[16] So it is reasonable to suppose that most of the money accounted for under the headings 'Kitchen Garden', 'Nurseries, Woods and Plantations' and similar descriptions were spent on the construction of new gardens and amenities. In any case, it is quite certain, even if the figure overstates the spending involved in the creation of entirely new work, that over £90,000 was spent on the house and park at Holkham during the time of their creation.

This took thirty-four years, from 1732 to 1765. Thus the average annual outlay was about £2,700. It is not always understood that building a house could take a very long time and that paying for it could therefore be attempted out of income.[17] However, spending

[15] M. Jourdain, *Work of William Kent* (London, 1948), 30-1 and 51; M. Brettingham, *The Plans Elevations and Sections of Holkham in Norfolk*, 2nd edn. (London, 1773), v. To Lord Hervey, Kent was a man 'with a very bad taste and little understanding, but had the good luck to make several people who had no taste or understanding of their own believe that they could borrow both of him, and had paid for their error by ruining their fortunes in making gardens and building houses that nobody could live in and everybody laughed at'. R. Sedgwick (ed.), *Lord Hervey's Memoirs*, ii (London, 1931), 581.

[16] See the inscription on the plaque inside the marble hall at Holkham: 'This Seat, on an open barren Estate, was planned, planted, built, decorated, and inhabited the middle of the XVIII century', and Historical MSS. Commission, 15th Report 6, Carlisle MSS., p. 86.

[17] The Wentworth Woodhouse papers show that the builder of that house left a larger estate and smaller debts than those he inherited. Even immense new houses did not inevitably bring debts or sales of land. For remarks on the possibility of building out of income, see H. J. Habakkuk, 'Daniel Finch, 2nd Earl of Nottingham: His House and Estate' in J. H. Plumb (ed.), *Studies in Social History*, 153 and 163-4.

year by year at Holkham was not regular: during the ten years 1732-41 about £20,000 was spent, in 1742-51 about £23,000, while in 1752-61, £42,000 was paid out, including £23,300 in the four years 1754-7, an average for those four years of £5,800 a year. This big spending was for finishing the central block of the house, in particular the extravagant and beautiful marble hall.[18]

Lord Leicester died in 1759 with his buildings incomplete. He laid down in his will[19] and made it one of the duties of the trustees to whom the estate was to be conveyed that they were to raise £2,000 per annum, free of deductions, and to use it to carry on 'the Building and finishing and compleating my said Capital House at Holkham and the offices Stables and Gardens thereto belonging until the same are fully and compleatly finished according to the Plan and Design which I have made, and which I have signed and approved of or which I shall have at the time of my Decease'. His wishes were carried out, and work went on after his death until 1765. Perhaps £2,000 was roughly what Lord Leicester thought a reasonable sum to spend annually on building. But he had not hesitated to spend far more in the years before his death. Probably he was anxious to see the house finished before he died; probably too, the death of his only son, Edward, in 1753, which meant that the estate would ultimately go to his nephew, caused him to feel less concern about the financial burdens he might pass on, or about the diminution of the estate that might be needed to make it possible to build fast. In a letter of 1754 in which he discussed sales—the proceeds to pay debt—he wrote 'my present income, which is what, having no son, I am chiefly to consider, will be much increased'.[20] In other words, he would have more to spend on his new buildings.[21] It was always possible, and sometimes imperative, for a landowner heavily burdened by debt to increase his disposable income by sales of land. The return from land was always smaller than that from a mortgage: it cost more to buy a given annual sum from land than it would to buy the same annual sum from a mortgage. By raising a capital sum by means of selling land an equal

[18] 'Country Accounts' book (Muniment Room).
[19] Dated 25 May 1756, Pr. 2 June 1759, H.F.D., 56B.
[20] C. W. James, op. cit., 268.
[21] His continued interest in them is shown in his letters to the duke of Newcastle, e.g. B.M. Add. MS. 32731, f. 13 and Add. MS. 32734, ff. 91-2, of 1754: 'I think of staying a few weeks longer to see finish't under my own eye, which alone I can trust, some principal parts here, which really raise so upon me, as to outdo my own expectations.'

capital value of mortgage debt could be discharged. The income lost through the sale of that land would be less than the amount of interest it would no longer be necessary to pay. The gain in immediately available income would be still greater if the value of the land to be sold lay partly in the reversionary profits to be expected from it as well as in its existing return. Sale of land, however, was usually a thing landowners were anxious to avoid. Their wealth and their prestige came from the land: mortgaging was a process by which a landowner could raise sums of money without permanently alienating part of his landed capital—and only if mortgaging or other borrowing was too freely indulged in or if an urgent need arose for more disposable income, would sales become unavoidable.

In spite of the steady increase in his income, and in spite of the contribution made by Lord Leicester's son to his father's finances, Lord Leicester was compelled to sell land in order to keep his debts within tolerable limits. His son, Edward Coke, made his contribution when he married Mary Campbell, one of four daughters of the duke of Argyll. Her portion was £12,500 and she was to have a share in the duke's personal estate after her mother's death. Edward and Mary passed all this on to Lord Leicester when they married in April 1747. In return, Lord Leicester promised to make provision for Edward and Mary in his own lifetime until Edward came into the estates, and for a jointure of £2,000 a year if Mary became a widow. At the time of the marriage, the duchess advanced £7,500 to Lord Leicester from his share of the duke's personalty. The duke left £67,021 personal estate (of which £50,600 was a mortgage debt on the estate of the late earl of Oxford).[22] Income from the Coke estates reached its highest level in Lord Leicester's time in 1749 when gross rents were £14,410. 10s. 5d. (supplemented that year by unusually high casual profits of £1,041. 15s. 0d.). In 1730 gross rents had been £11,565. 4s. 4d.; in 1740 £12,629. 3s. 5d. After 1749 sales brought down the total income from rents, and in 1755 they were down to £12,095. 17s. 8d., but by the time of Lord Leicester's death in 1759, assisted by purchases, they had climbed

[22] H.F.D., 51. The 2nd duke of Argyll died in 1743. The duchess died in 1767. Mary's share in the duke's personalty was just over £14,000. The marriage was unusually mercenary even by contemporary standards. According to Horace Walpole (letter to George Montagu, 3 July 1746), Edward Coke was first offered to 'all the great lumps of gold in all the alleys of the City'. It was one of the most disastrous marriages of the century; there was never, it seems, any question of an heir.

back to £13,665. 0s. 6d. Net income from the estates averaged £8,489. 19s. 1d. in 1730–5; in 1753–8 it was £9,964. 13s. 1d.[23] Very substantial sales took place between 1749 and 1756. Knightley in Staffordshire went in 1750, the Coke estates in Suffolk, together with Farnham Royal in Buckinghamshire, followed in 1752, and the Dorset estates (Donyat, Shillington, and Durweston) were sold in 1755. The lands sold had brought in gross yearly rents of over £3,000. All of these sales were of lands outside Norfolk, outlying estates, away from the nucleus of the 'Grand Estate'.[24] A major portion of these sales (of lands worth £1,775. 6s. 6d. gross in 1751) was made in 1752 in order to pay off a mortgage debt of £48,000 due to Sir Matthew Lamb which had been made in June 1741, probably to re-secure earlier loans from the Lambs as well as to secure new ones.[25] Some small sales had evidently been made in earlier years to meet current expenditure: in 1742 Lord Lovell wrote to Lamb 'I hope you will be able to sell the little thing in Suffolk and return us the money to live upon'.[26]

On the other hand, purchases of land in Norfolk continued and were particularly substantial in the last, financially difficult, years of Lord Leicester's life. An important purchase took place shortly before Lord Leicester's death of estates at Wighton in Norfolk to the value of nearly £14,000.[27] Other Norfolk lands were also bought in the closing years of Lord Leicester's life; in the Burnhams, in two transactions, in 1756 and 1757, for £4,130, in Dunton in 1754 for £2,500, in Fulmodestone for £2,200 in 1757, among others.[28] Mortgages had to be given to some of the vendors in lieu of some or all of the price. In the fifties, buying and selling of land, repayment of debt and new borrowing were going on practically simultaneously. The effect of the land transactions was to concentrate the Coke estates, to make them more largely a Norfolk estate. Lord Leicester seems to have tried, for a while, to reduce his debts—an attempt which soon came to an end, as a result, perhaps, of the death of his son in 1753.

At the end of his life, Lord Leicester left debts of £90,973. 18s. 10d. Of these, £60,357. 0s. 0d. were secured by mortgages, while £30,616. 18s. 10d. were debts on bond, or simple contract above

[23] A/B 1730–59. These figures do not include income from the lighthouse or from Lord Leicester's official salaries, etc.
[24] A/B 1749–56.   [25] H.F.D., 131.   [26] C. W. James, op. cit. 268.
[27] P.R.O., C. 12/572/13, R. Cauldwell's 'Answer'.
[28] Burnham Deeds 218, 220; Dunton 83, Fulmodestone 44.

what could be discharged from Lord Leicester's personal estate—which seems, indeed, only to have accounted for £200 of debt.[29] Lord Leicester provided in his will that the amount needed to pay off debts unsecured by mortgage should be raised after his death by a new mortgage. This was done, and lands in Norfolk were mortgaged on 20 May 1765 to Charles Yorke and others. The interest was to be 4½ per cent, or 5 per cent if the payment of interest were left in arrears for four months or more.[30] Lord Leicester's will also directed that £3,000 a year was to be devoted to paying off his debts until all of them were discharged, but he added that this sinking-fund was only to come into operation after the death of Lady Leicester; the estates were exempted from the burden of repayment while she lived. This suggests that Lord Leicester was anxious not to burden his closest kin too heavily and makes it seem more likely that the death of his son in 1753 encouraged him to be extravagant. After his death, his estates were to go to his wife, but after her death they would pass to a nephew, Wenman Roberts. It was on this nephew and his descendants that the burden of Lord Leicester's debts was laid.[31]

Of the mortgage debts £31,457 was owed to Sir Matthew Lamb. This was in three mortgages: one for £16,000 at 4 per cent made in March 1752; one for £5,457 at 4 per cent made in July 1741 and the third made in March 1757 for £10,000. £6,000 was owed to the trustees in the will of the earl of Thanet—this was the remainder of £10,500 borrowed at 5 per cent from Hugh Chamberlen in 1720. £7,000 at 5 per cent was still due, as it had been since 1716, to Philip Roberts, or rather to his trustees. £8,000 was due to Mrs. Mary Tufton, debt contracted at 4 per cent in May 1750. £3,000 was due to Dr. Thurston which had been borrowed from him in 1756 at the remarkably low interest rate of 3½ per cent. This was evidently to pay for lands bought from him. £1,800 was due to William Fellows Esq. borrowed at an unknown date. £2,200 was due to Mr. Francis Hill; this was contracted in October 1757 when lands were bought from Hill and mortgaged to him, i.e. the equity of redemption was bought. £900 was due to Mrs. Hase—in this instance Lord Leicester had bought the equity of redemption of some land in Wighton from a previous mortgagor.[32]

---

[29] H.F.D., 62.   [30] H.F.D., 57.   [31] H.F.D., 56B.
[32] H.F.D., 48, 54, 57, 59, 68 (mortgages or assignments), Burnham Deeds, 218, 220 (Conveyance and Mortgage). G.E.D., 78 (reconveyance of trust by Cauldwell). The equity

The possibility of contracting mortgages on the scale he did was open to Lord Leicester because, when his estates were settled, there were left out of settlement substantial portions of his estate, and later, the whole of his estates escaped from settlement altogether (hence the power he had to dictate their destiny by will). Had the facts been different, Lord Leicester's borrowings on mortgage might have been impossible, for settlements prevented the borrowing of money on the security of the land settled except for purposes specified in the deed—usually only the endowment of younger children. At the settlement made when Thomas Coke married Margaret Tufton in 1718, the lands settled were the Norfolk estates. The estate outside Norfolk, which was substantial (it was worth £5,316 a year in 1739), was left outside the settlement so that it could be freely pledged as security.[33] When a new settlement was made on the marriage of Lord Leicester's son to Mary Campbell in 1747, again only Norfolk lands were settled, and some in Norfolk were left unsettled.[34] After the death of Lord Leicester's son, the estate escaped from settlement altogether.[35]

The borrowing operations of Lord Leicester and his advisers offer some features of interest. There seem to have been three levels of borrowing. There were the large sums, usually borrowed on mortgage security. These large loans were normally arranged by the Lambs and the money itself usually came from the Lambs (perhaps their own, perhaps money invested for some of their other clients) or from friends or family connections of the Cokes. The Lambs also provided short-term overdraft moneys, so to speak, at least in the 1720s. At that time we find payments to Lamb 'for Interest of Money advanced in the time of this Account', that is, during the course of a particular year.[36] Then, secondly, there were smaller but still quite substantial sums (one or two thousands) borrowed sometimes on mortgage, more often on less formal security, frequently in or around Holkham. Thirdly, there were small sums, sometimes as little as five or ten pounds, borrowed from

---

of redemption is possessed by a person possessing the right, on repaying capital secured on land, to gain or regain full ownership of that land.

[33] Holkham Lib. MS. 742. Copy of the settlement.
[34] H.F.D., 52. (Printed Act of Parliament.)
[35] G.E.D., 78. For the effects of marriage settlements, see H. J. Habakkuk, 'Marriage Settlements in the Eighteenth Century', *Transactions of the Royal Historical Society*, iv. 32 (1950), 15-30.
[36] Holkham Lib., MS. 741.

people in and around Holkham or, in the 1720s, borrowed through Lamb. It is possible, but no more than possible, that these small local borrowings became more prominent at times when it might be difficult to borrow in London, either because of a rise in the yield of investment in the public funds or because of a shortage of land out of settlement available for mortgage. There is no trace of any large mortgage being concluded during years of national financial stringency (it is probably significant that the £30,616 mortgage to Yorke was not settled until 1765, six years after Lord Leicester's death in 1759), but this may be mere coincidence. In any case, it was always worthwhile to avoid the legal expenses of contracting a mortgage.

Interest rates on mortgages tended to fall in the first half of the century, a fall interrupted by the periods of war. Before 1720 and at the time of the liquidation of the South Sea disaster, money was borrowed at 5 per cent.[37] By 1725, when £34,000 was re-secured,[38] the rate had fallen to $4\frac{1}{2}$ per cent, though some other mortgage debts remained at 5 per cent after that time.[39] In 1741 mortgages were made at 4 per cent[40] and in March 1752 a mortgage was arranged with Lamb for £16,000 at 4 per cent, if the interest were paid within four months of the date due, and 5 per cent if the interest were paid later.[41] In 1756 a mortgage was made at the lowest rate recorded until the 1830s, for £3,000 at $3\frac{1}{2}$ per cent.

Interest rates on non-mortgage debts were about the same as they were on mortgages, or perhaps rather higher. In the 1720s, even in the later years, it was 5 per cent, which continued to be paid in the 1730s. In the 1740s 4 per cent to 5 per cent seems to have prevailed, sometimes 4 sometimes 5. In the 1750s the rate tended to fall, and by Lord Leicester's death in 1759, only few non-mortgage debts existed at more than 4 per cent interest.[42]

From what sort of people did money come? In the South Sea period the large mortgage loans came from Lamb or from those who had sold stock to Coke, who seem to have been friends or family connections or London financiers and brokers. Lord Finch for instance, who gained at Coke's expense at the time of the South

[37] H.F.D., 39, 44 (ii); Holkham Lib. MS., 'Ledger book of Mr. Coke's Accounts'.
[38] H.F.D., 44 (iii).
[39] Holkham Lib. MS. 741.
[40] H.F.D., 50, 63.
[41] H.F.D., 60; A/B 1756.
[42] Holkham Lib. MS. 741; A/B accounts current.

Sea crisis, and who lent to him on mortgage afterwards, was a familiar figure at the gaming tables frequented by Coke and lost £1,050 to him at play in 1722-3. Later, such loans came almost entirely from Lamb or from people from whom land was bought. The larger lenders on non-mortgage security were often of the same type as the mortgagees. But local men often lent such sums as well as the rich friends, London financiers, and clients of Lamb, who were still, in the 1720s, the main lenders in this department. Coke borrowed £2,000 from his brother Edward in 1726[43] and borrowed £3,500 from his nephew, Wenman Roberts, some time before his death.[44] He borrowed £2,000 from Henry Pelham on bond on 4 October 1726 and repaid him on 2 December 1726.[45] An account of this loan is contained in a letter of Matthew Lamb's of 11 October:

> Sir Thos. came on purpose [to London] to gett the money to satisfie the Dutchesse of Buckingham demands, & as soon as he came he told me he didn't know what he cou'd possibly do for the money unlesse you wou'd be so obliging to lend it him, but being in Company with Lord Sunderland and Mr. Pelham, he was telling how the Dutchesse of Buckingham had served him, & Mr. Pelham said it was very Scandalous & that he had 2000£ which was at Sr. Thomas's Service, which Sr. Thomas borrowed of him upon his Bond, and as to the other 1000£ Sr. Thos. told me he wou'd write to you . . .[46]

Mr. Waller, the London financier, lent Coke £2,000 on bond in 1726. Such transactions were frequent. The most striking example of similar loans from local people was in 1757, when Benoni Mallett, Henry Savory, and Nathaniel Cowper each lent £1,000 on bond with interest at 4 per cent. All these were local men; Benoni Mallett held a large farm at Dunton as Lord Leicester's tenant.[47] William Collison lent various sums on note in 1745 and 1746, totalling £305 at 5 per cent. The money was paid back in 1755.[48] Collison was in the beer trade. He was tenant of alehouses in Flitcham, and in Tittleshall, where he also held a malthouse, a brew office, a house, and some land.[49] Several sums of money were borrowed at various times on bond and notes from Henry Knatts. By 1742 he had lent £364 and in April 1742 he lent another £105 followed by £70 in

[43] Holkham Lib. MS. 741, f. 26ʳ. [44] H.F.D., 62.
[45] Holkham Lib. MS. 741, ff. 29ᵛ and 31ᵛ.
[46] To an unknown correspondent 11 Oct. 1726. Holkham Lib. MS. 728, iv, f. 198.
[47] On 21-year lease from 1748 at £225 per annum. A/B 1757.
[48] 'Accounts of Debtor and Creditor' (Game Larder). [49] A/B 1750.

November and £81. 12s. 8d. in December 1743. In February 1745 he lent another £11. 11s. 6d. and £40 in 1746. No more was had from him until January 1749, when he lent on note another £44. 13s. 1d. Then in 1750 he lent £300, took over debts of £161. 19s. 0d. due to other people, and took a bond for his bill for oats of £44. 6s. 8d. By January 1750 Knatts was owed no less than £1,314. 15s. 7d. At that point, mysteriously, £24. 15s. 7d. was repaid of the principal and Lord Leicester gave his bond for the remaining £1,290 with interest at 4½ per cent. This was still owing in 1759, and in the meantime Knatts had lent another £500 which was soon repaid. Henry Knatts was the tenant of a farm in Holkham, for which he was paying £118. 10s. 0d. annual rent in 1751. The tenant of the Beckhall farm (£165 per annum), in Billingford, lent £80. 11s. 1d. in November 1748 which was repaid in July 1751, and £44. 14s. 2d. in November 1753, which was repaid in 1754. In 1755 William Elliott of the Branthill farm at Holkham, was repaid £400 at 4 per cent, and in the same year Thomas Tann, another Holkham tenant, got back £480 at 4 per cent.[50]

Another prominent lender was a local man in a different position, Daniel Jones, an attorney in Fakenham. Between 1742 and 1744 he lent £1,033. 10s. 0d. on note in eleven instalments, the largest of which was £300, the smallest £20. This was paid early in 1745. In 1745 and 1746 an account was kept going with him in which repayments followed fairly quickly on borrowings, but in 1748 to 1751 quite substantial borrowings took place and repayments fell far behind. One sum borrowed was for a bill for legal charges of £138. 8s. 1d. and another sum of £104. 12s. 5d. looks like a bill. The rest were evidently pure loans. In early 1752 £1,660 was paid to discharge the debts due to Jones. Then the account was closed.[51]

In some ways those who lent really small sums are the strangest phenomenon. Among the debts outstanding at Lord Leicester's death were eleven of less than £100 including five for less than £50.[52] The lenders were people like Michael Pattern, to whom £180, at 4 per cent, was repaid before 1759. When he gave a small loan in 1751, he was noted as being a 'Husbandry Ser't'—he was an employee of Lord Leicester's. In 1751 £80 was borrowed on note from Thomas Binney, the Holkham shepherd, and by 1755,

---

[50] 'Accounts of Debtor and Creditor', A/B 1740-54; A/B 1755 (accounts current); H.F.D., 62.
[51] 'Accounts of Debtor and Creditor.'   [52] H.F.D., 62.

General Finance, 1722–1759 33

£140 at 4 per cent was due to him. On 18 July 1750 John Elliott, a bricklayer, lent on bond no less than £400 at 4 per cent.[53] Earlier, in the 1720s, money was raised from similar sorts of people. On 14 April 1728 a farrier, Mr. Dark, was repaid £59. 11s. 0d. with interest for nearly two years and Mr. Huddleston, a mercer, was repaid £39. 18s. 7d.[54] Evidently these were unpaid bills for work done or goods supplied, and many of the debts contracted from time to time by Coke's stewards and agents were of this kind. Possibly the debts to servants represented unpaid wages—perhaps the servants were anxious to save in this way. But many small debts, especially the ones that were not among the very smallest, seem to have been the result of genuine loans, when money the lenders had accumulated was lent to the Coke steward when he needed money to pay bills or wages. The steward on occasions, might not despise quite small sums.

These relatively small non-mortgage loans were evidently intended to be temporary, to tide the estate over brief periods of shortage of money; but such loans were liable to become permanencies—as at Lord Leicester's death when £30,616 of them were paid off only by raising a loan on mortgage. Tradesmen probably often had to take a security instead of payment of their bills: the risk of such delays was a normal hazard of eighteenth-century dealings with the great. But apart from them, there seem to have existed in the hands of Coke tenants and other inhabitants of northwest Norfolk small savings which they were ready to lend on reasonable security—and lending to landlords like Coke was safe enough in the long run; signs of strain on his finances were not at all signs of impending ruin. That very humble people had sometimes fairly large sums to lend is shown by inventories made of the personal property of deceased villagers. For instance, when Edward King, a yeoman of Massingham, died in 1724, his entire movable wealth was £625. 1s. 8d. Of this £50 was cash in hand and he had no less than £517. 19s. 8d. out on loan.[55] For Coke and his stewards in the early eighteenth century the inhabitants of the neighbourhood collectively supplied the place of a bank overdraft,[56] and Coke provided a savings bank service for dwellers around Holkham—a savings

[53] A/B 1750–9 (accounts current).
[54] Holkham Lib. MS. 741, f. 40ᵛ.
[55] Norwich Castle Muniment Room, Case 33, Shelf F/17, no. 54.
[56] Lord Leicester was in credit at Childs' when he died, with £627. 13s. 11d. there. H.F.D., 62. Miss Ashbee, archivist to Messrs. Williams and Glyns, tells me that no records survive of his dealings with Childs'.

bank of a rather primitive kind, one not always ready to borrow and often far from ready to repay, but a precursor of the country banks of the later half of the century.

In spite of these transactions, Lord Leicester was consistently in debt to his own stewards. This debt arose as a result of the excess payments made by the steward over net receipts for rent. The bulk of those payments was money returned to the banks in London. The effect was that the stewards provided an overdraft facility. In return they were allowed interest at 5 per cent on the balance due at the closing of the previous annual audit account. In the 1730s these balances fluctuated between £450 and £1,680 (at 15 September 1740). In June 1741 Appleyard, the steward, received this sum in full from Lamb with £64 interest. However, by the end of the 1740s, Cauldwell was owed over £1,000 and by the middle of the 1750s the balance due to him had reached £2,000. It would be interesting to know how the stewards financed these loans. It may be suspected that the interest credited to them represented a valuable perquisite since manipulation of the accounts to maximize debit balances at the closing of the books should not have been impossible. Such current account borrowing from stewards was unknown in Coke of Norfolk's time.[57]

From 1730 (and this may explain the decision to start building) Coke finances were fortified by the revenue from the most striking and valuable benefit that Thomas Coke gained for himself and his descendants by marrying Margaret Tufton: the lease from the Crown of the Dungeness lighthouse. As part of Margaret's fortune, her father, the earl of Thanet, assigned his grant of the lighthouse to Lord Lovell, and Lord Lovell took up Thanet's grant when he died in 1729. The holder of the lease was to pay to the Crown a yearly rent of £6. 13s. 4d. and promised to maintain a light for the 'safety and direction' of mariners. He was to renew the light or change its position according to the directions of Trinity House 'or of other expert and skilful Seafaring men'. In return the holder of the grant was given power to take one penny per ton from masters of all ships, British and foreign, inward and outward bound, who passed the light. Strictly speaking, the lease was of the right to collect these tolls, the land on which the lighthouse stood was not Crown property nor was the lighthouse erected by the Crown.[58]

[57] A/B 1722–56 (accounts current).
[58] B.M. Add. MS. 36128, f. 325, copy of warrant for letters patent. This says that charges

The original grant was a characteristic product of the seventeenth century, when monarchs, always short of ready money, rewarded servants, repaid debts and—in Charles II's case—endowed their children, by, in effect, delegating to them the right to receive revenues. The king gave these people an income by giving them an opportunity to levy tribute on the public, instead of producing a capital sum for them himself. Such grants sometimes went on for a very long time; once given, they soon came to be thought of as ordinary property, not to be lightly impugned, and their possession implied no particular political attitude in the beneficiary. As late as 1815, the Lords of the Treasury informed Harwich corporation that preference was normally given to ancient lessees of the Crown when leases came up for renewal and that therefore the Harwich lights would remain in private hands. Grants to private persons of the right to establish lighthouses began in James I's reign. Trinity House had been given the right to erect seamarks in 1566 but in 1617 the Attorney-General advised that this did not prevent the Crown from doing so—this right of the Crown could be delegated or sold. Most of the lights established in the seventeenth century, when the modern English lighthouse system began, were constructed by private individuals, in spite of the constant objections made by Trinity House to private property in lights, and after the Restoration, applications for grants became very frequent. Sir John Clayton, according to Pepys, received no fewer than five lighthouse patents in 1677.[59]

Charles II in 1680 gave Richard Tufton, later earl of Thanet, a grant of the light at Dungeness, a grant to come into effect when an existing grant to one Elizabeth Shipman should have expired, and to last for thirty-one years. Elizabeth Shipman's term evidently expired in 1706. Thus Thanet's grant, which Lord Lovell took over, was due to end in 1737, and it seems that Lord Lovell's grant was then renewed for another thirty-one years, to last until 1768. In 1740, however, Lord Lovell managed to secure yet another grant to run for sixty years from the end of the one he held then, that is, until 1828. It is not certain why he was so fortunate; probably his political

were to be on inward bound ships, but all other evidence about the lighthouse refers to tolls on ships inward and outward bound and the warrant may have been mistranscribed.

[59] For the history of private lights see W. R. Chaplin, 'The History of Harwich Lights and Their Owners', *The American Neptune*, xi (1951), 5-34; and 'The History of Flat Holm Lighthouse', ibid., xx (1960), 5-43.

influence in Norfolk was important enough for the government to be concerned to oblige him. There is no doubt that the pretext alleged—the cost of building a new lighthouse—was a barely plausible excuse for his request. A new lighthouse would not cost more than could be paid out of the ordinary revenue from the tolls for one or two years. In any case, the consequence was that the private light remained in the possession of the Cokes until a time when political opinion was much less sympathetically inclined to that sort of property.[60]

It was an extremely profitable concern, more and more so as time went on. Early in the Coke tenure, in the period from Christmas 1730 to Christmas 1732, gross receipts—receipts coming in to the agents at London, after deduction of commission at the ports—were £3,298. Expenses—salaries to lightkeepers, upkeep of the light and lighthouse, salary to the supervisor at Lydd, and London agents' fees—were £549. Net receipts, the income to Lord Lovell, were therefore £2,749, equal to about £1,375 each year. As more shipping moved up and down the Channel, income increased; in 1760, soon after Lord Leicester's death, the income from the light was just over £2,000.[61]

In this way the Cokes shared in the advancing commercial prosperity of Britain and Europe. Unlike more fortunately situated landowners, this was the only way in which they did so. Their London property, Bevis Marks, was never very important—it brought in only £200 a year until it was sold for £9,250 in 1787. They sold nearly all the coal they owned without comprehending, until too late, its potential value. The lighthouse gave them a share in European commercial prosperity—they were unlucky that this mark of royal support came in this, as it proved, easily attacked form. On the other hand, the Grand Estate, the main Coke landed property in Norfolk, was highly improvable.

[60] Similarly, Sir Isaac Rebow was given, in March 1716, a grant of the Harwich lights for sixty-one years from the expiry of his existing grant in 1756 (W. R. Chaplin, 'Harwich Lights', 19). See Appendix 1 for details of the management of the lighthouse.
[61] Box of lighthouse accounts in game larder.

4

The estates under Thomas Coke first earl of Leicester

i. *The size and value of the estates*

WHEN Thomas Coke came of age, in 1718, his estates lay in Norfolk, Suffolk, Kent, Buckinghamshire, Oxfordshire, Staffordshire, Somerset, Dorset, and London, including those still in jointure to his grandmother, and brought in, from gross rents, plus average annual casual profits, just over £10,000 a year. Improved rents in those estates where lifeholds were common were calculated to be worth, in reversion, an extra £3,600 per annum. Twenty church livings belonged to the estate. The lands in Norfolk brought in about £6,120 a year in gross rents and estimated casual profits.[1] Lady Anne's jointure lands were leased from her for £1,200 a year —the land thus restored to the Grand Estate brought in an average revenue, after deduction of taxes and all other reprises, of about £1,360 per annum.[2]

In 1758, the last full year before Lord Leicester's death,[3] the estates produced gross rents of about £13,500.[4] At its largest and most productive, in Lord Leicester's time, in 1749, the estate brought in gross rents of over £14,400 plus £1,010 in casual profits, a total of over £15,000. Of this, £9,100 came in from gross rents in Norfolk (plus £238 casual profits) and £5,300 from estates elsewhere (plus £800 casual profits).[5] After then, while Norfolk rents continued to increase, and reached over £10,500 per annum by 1758, the rents received from other estates declined as sales of land reduced their size.[6] The effect of these sales, added to purchases of

---
[1] Holkham Lib. MS. 741, f. 1ᵛ. Casual profits in Norfolk were small—about £100 per annum; there, rack-rents, as distinct from rents reduced by capital payments at the making of leases, were the rule.
[2] In Holkham Lib., 'Estate Accounts, 1679-1709'; H.F.D., 28B and 43.
[3] Once again, the reader is reminded that Coke became Baron Lovell in 1728 and earl of Leicester in 1744.
[4] A/B 1758.   [5] A/B 1749.   [6] A/B 1749-58.

Norfolk lands, was to concentrate the Coke estate, which came more and more to be a Norfolk estate. Between 1720 and 1749 gross annual rents in Norfolk rose by 47 per cent,[7] and gross annual rents in the other estates by 83 per cent.[8] In the estates outside Norfolk the increase was made greater by the replacement of tenancies at low rents after high entry fines by tenancies at full annual rents. Between 1720 and 1757 gross annual rents in Norfolk rose slightly more than 70 per cent.[9]

Much of this substantial increase in Norfolk rents was due to the purchase of new estates. Between 1718 and 1746 there were bought, in Norfolk, lands of the yearly value, when purchased, of £1,070. These were brought into settlement in 1747 (when Lord Leicester's only child, Lord Coke, married Lady Mary Campbell) in exchange for lands, which were then released from settlement, in outlying parts of Norfolk, detached from the main estates. The lands bought were all small parcels lying 'intermixed with, or contiguous to, several of the Farms and Lands comprised in the said Settlement, and are now lett without Distinction together with the same; by which means, the yearly Value of the said settled Estate is greatly improved and increased'.[10] The largest single item (apart from the manor and demesnes of Quarles, worth £164 per annum, leased from Christ's College, Cambridge, from 1719) was several messuages, cottages, and lands in Tittleshall of £136 annual value bought from John Carr, in 1734. In nearly every year something was bought to add to one part or another of the Norfolk estate.[11] Quite clearly, it was an established policy to buy, whenever they came up for sale, properties which might round off the main Norfolk estates. Such lands were apparently bought without question, whatever the financial condition of the family. Immense debts, and extravagant building costs, never deterred Lord Leicester and his advisers from making this sort of purchase. The most striking example is from 1758, when £9,500 was paid for an estate in Wighton.[12] This was one of a number of purchases in Wighton made just before Lord Leicester's death: in all, nearly £14,000 was paid

[7] £6,192 to £9,094.
[8] £2,910 to £5,316.
[9] £6,192 to £10,541.
[10] H.F.D., 52: Printed *Act for settling the Estate, of Thomas Earl of Leicester and Edward Coke Esq, commonly called Lord Coke, his only son, on his marriage with Lady Mary Campbell, one of the daughters of John, late Duke of Argyll, deceased.*
[11] G.E.D., 54.   [12] Wighton Deeds, 272. From Christopher Bedingfield.

The estates under Thomas Coke, first earl of Leicester

for them.[13] It was in the fifties, when these estates were bought, that spending on building was at its height. Such purchases of land might involve selling land away from the main estates but this did not prevent them. The preservation, development, and good management of the main Norfolk estates was always taken for granted, and nothing short of total financial catastrophe would have been allowed to hinder their progress. These purchases were designed to round off the existing estate, to make its management simpler and more efficient; they were certainly not a consequence of the possession by the landlord of idle sums of capital which he had ready to invest in any remunerative way.

Gross rents in Norfolk rose from £6,016 in 1718 to £8,616 in 1746.[14] We have seen that lands worth, when purchased, £1,070 a year were added in these years. This implies an increase in rents during these years of about 25 per cent due to management alone. Rents continued to increase in Norfolk in the 1750s, and by 1759 they were equal to £11,153 per annum.[15] According to Ralph Cauldwell, he bought lands in Norfolk, while he was steward under Lord Leicester (1742 to 1759), which were worth—by 1759—£1,812 per annum. During this period annual rents in Norfolk rose from £7,866 to £11,153.[16] Making allowance for the amount Cauldwell ascribed to his purchases it seems that rents for a constant amount of Norfolk land rose in the years 1742-59 by about 20 per cent, a rise for which Cauldwell claimed credit, due to management alone.

It is possible to work out an approximate percentage increase in rents over a constant area of land for the years 1718-59, the years of Lord Leicester's adult life. The figure is 44 per cent.[17]

ii. *The improvement of the estates*

What caused this increase? There was certainly no general upward movement in prices over the forty years in question to explain the rise in rents. Thus the rise in rents implied a rise in real income and it was not due to rising prices for farm produce. There must have

---

[13] P.R.O., C. 12/572/13. R. Cauldwell's answer to Bill in Chancery, 20 Nov. 1778.

[14] Figure from 1718 from Holkham Lib. MS. 741, f. 55^v where Lord Lovell himself wrote down a comparison of his rental income in 1718 with that in 1739. A/B 1746.

[15] A/B 1759, figures for half year to Lady Day.

[16] A/B 1742, 1759. Cauldwell gave the figure for 1742 as £7,712. 7s. 8d. P.R.O., C. 12/572/13, ff. 3-4.

[17] In these calculations it is assumed that newly purchased land rose in value per annum at the same rate as land already owned.

been a growth in the productivity of land held by the farmers in Norfolk. It should be noted, however, that the landlord's net income did not rise with the smoothness and regularity of the gross rental. The 1730s and 1740s were difficult years for farmers, with abundant harvests and low prices. Thus in 1736 arrears in rent payments of over £3,000 were recorded. Still, by the 1750s, the level of arrears was back to that of the beginning of Lord Leicester's period in control of his estates and it is reasonable, therefore, to take the increase in gross rents as indicating an underlying agricultural progress.

An increase in productivity came about through the more intensive use of farming land. The application of capital and the deployment of new crops made this possible. The history of agrarian progress on the Coke estates in the eighteenth century is continuous; there was no revolutionary upheaval. The fact that crops of comparatively recent introduction were in use under the guardianship has already been illustrated. Such crops must have brought increased production on land already frequently cropped; furthermore, they contributed decisively to making it possible to bring land hitherto uncropped, or comparatively rarely cropped, into fuller cultivation. Intensive cultivation was being extended, both inward, towards the centre of villages as well as outward towards the periphery of villages; open-field strips at the centre of villages were disappearing and their disappearance allowed the land to be used more effectively; at the same time land once given over wholly or partly to sheep pasture or waste came into full cultivation. To farm the light Norfolk soil needed capital and skill and they were provided both by the landlord and by the tenants.

Thomas Coke contributed to the expense of enclosure and marling, encouraged skilful tenants by giving them long leases and good farm buildings, and even sought to compel them, if compulsion was needed, to use up-to-date crop rotations.

The advantages of enclosure have often been discussed, usually in terms of the liberation of one cultivator from the need to consult others about what he should do. Thus the extinction of rights of commonage has become comprised in the word 'enclosure'. The little-cultivated external lands of villages in north-western Norfolk were often subject to rights of sheep-walk—these had to be extinguished before a cultivator could be free to improve and experiment. In north-west Norfolk, fences, above all, kept other people's sheep out, or one's own in, both on former strips and on land formerly given over to sheep pasturage. On strip land the erection of fences made it possible to substitute crops for a fallow without other people's agreement while on former sheep-walk fences were a prerequisite of any cultivation at all. Permanent fences, literal enclosures, were better than temporary fences because they saved labour.

The landlord made enclosure easier by acquiring small tenements intermixed with, or adjacent to, existing farms, by buying or leasing rights of pasturage, and by contributing to the cost of erecting fences. Lord Lovell, for instance, promised William Lee, tenant of Longlands farm, Holkham, to 'inclose and divide into proper Inclosures as was by them Verbally agreed all those lands at p'sent uninclosed in the space of 2 years'.[18] In the first half of the eighteenth century the estate account books contain a very large number of references to payments allowed to tenants for the construction of fences. In 1736, for example, Carr of Massingham was allowed £18 for 6 cords of battens for his new enclosures.[19] By 1752 John Carr could point out that there was on his farm only one unenclosed field.[20] Just as the policy of buying up small tenements was one continued from the guardians' time, so too there was nothing new about enclosing. The previous Carr was allowed £17 in 1712 for 'making gates and setting down posts for the new inclosures to be taken in', and the description of the Wicken farm, Castleacre, in 1716, mentions 'new inclosures' of over 62 acres.[21]

Similarly, payments for marling, familiar in the guardians' time, were continued by their former ward. The clearest evidence of the way in which this was done is in the agreement of 1733 between Lord Lovell and the new tenant of Longlands farm, William Lee.[22] Lee was to pay 9s. per acre yearly rent for land already 'Improved & Marl'd' and 8s. 6d. per acre for lands after they had subsequently been thus improved. Until the unmarled lands were marled Lord Lovell agreed to allow for them 2s. 6d. per acre abatement in rent, so that the rent per acre was only 6s. before marling. Lee was to arrange for all the unmarled lands to be marled within four years. He was to find carriage for the marl, but Lord Lovell promised to

[18] Holkham Deeds, 1072.
[19] A/B 1736.
[20] In Massingham Deeds, 236. Letter of John Carr to Cauldwell, 18 May: 'you know there is but one Field uninclos'd.'
[21] A/B 1712, 1716.
[22] Holkham Deeds, 1072.

allow him, to be deducted from his rent, 27s. per hundred loads for digging, filling, and spreading the marl. Thus for a farm for which £138. 8s. 0d. was paid by Lee in 1734, he paid £148. 5s. 6d. in 1735, £9. 17s. 6d. being added to the rent for 79 acres of land newly marled. In 1736 he was holding his farm (with extra lands added since 1733) at £230, reduced by £22. 3s. 0d. to £207. 17s. 0d. on account of lands still unmarled. By Michaelmas 1737, after another 4,060 loads of marl had been spread, the rent was £220 (no reason is given for the reduction from £230). By 1745 the farm was let to another tenant at a uniform rent of 8s. 3d. per acre for 446 acres—evidently the farm had been reduced in size since 1737.[23]

Presumably under a similar contract, William Dewing of South Creake was allowed £16. 18s. 0d. in 1727 for spreading 1,300 loads of marl at 26s. per hundred; this was part of 5,000 loads he had agreed to spread in three years. In 1748 'money spent on improvement' included £70 allowed to William Kent at Weasenham for carrying 5,856 loads of mark at £12 per thousand; at Fulmodestone, John Drosier was allowed £11. 5s. 0d. for 1,000 loads of marl; and in Holkham, John Elliott, of Branthill farm, was allowed £30 for 2,000 loads of marl.[24] These and other references make it clear that marling was taking place in one part or another of the estate in Norfolk throughout the first half of the eighteenth century, and that the landlord was assuming a major share of the cost. It is not surprising to find one of the enclosures on Carr of Massingham's farm changing its name in this period from the infertile sounding 'Braky Break' to that of 'Marl Break' or 'New Marl'd Break'.[25]

Marling provided a major weapon in the advance of neat permanent enclosures, the setting out of fields soon to be capable of high cultivation, an advance that moved inwards, abolishing strips, and outwards, dividing up sheep pastures. The stages in the process can be seen by studying the estate maps and by considering the evidence of increasingly intensive cultivation. In 1779 there were still some town commons left and in some parishes a few strips but very few acres of mere sheep-walk.[26] Earlier, in the 1750s, there

[23] A/B 1734-45.
[24] A/B 1727, 1748.
[25] Massingham Maps 5/85 (after 1728), 5/86 (after 1744), and 'Plans' map (1779).
[26] Use of the former sheep-walks had often been limited to fewer persons than that of commons properly so called. Indeed, great tracts of sheep-walk were often let with farms to one tenant.

The estates under Thomas Coke, first earl of Leicester 43

were more strips, commons, and sheep-walk, but it is clear that their area had been much reduced since the earlier decades of the century. The pattern is not constant and the chronology varies. Some areas were almost entirely enclosed by the beginning of the eighteenth century. Waterden, for instance, in 1714, had no waste land, sheep-walk, or commons, and only vestigial strips—which had gone by 1789.[27] Other villages had large areas of strips early in the eighteenth century, which were greatly reduced in the first three-quarters of the century.[28] Maps of Tittleshall show that some land lying in strips was enclosed during the seventeenth century, but a substantially larger amount of enclosure took place between 1725 and 1779, both of strips in the centre of the parish, which disappeared altogether, and of areas of sheep-walk or waste on the periphery.[29] Fulmodestone had no strips as early as 1614, but Fulmodestone Common, of over 500 acres, survived beyond 1789.[30] In Longham there were many strips around 1580; before the first quarter of the eighteenth century, consolidation had taken place and some strips had vanished into enclosures: Mapp's Yard, for instance, a close of 20 acres in the early eighteenth-century map, was in twenty-three strips in the sixteenth-century one.[31]

This discussion of field structure in the early eighteenth century shows the danger of basing any far-reaching inferences about early field structure on the existence, or non-existence, of late eighteenth- or nineteenth-century enclosure acts. Castleacre and Tittleshall had no enclosure acts while Fulmodestone, for instance, did, yet the latter parish had no strips in the eighteenth century, while Castleacre and Tittleshall both did.[32] It is incorrect to assume that a late eighteenth-century or nineteenth-century enclosure act implies that little enclosing had taken place in the relevant area earlier on. Small landowners were bought out before, as well as after, enclosure, common rights could be bought up, and owners encouraged to make agreements for partial enclosures. Subsequent enclosure acts sometimes had a purely marginal effect.

Enclosures of former strips brought about after exchange or

[27] Maps 3/48 and 3/49.
[28] e.g. Castleacre Maps 5/80, 5/81, and 'Plans'.
[29] Maps 4/73a, 4/74, and 'Plans'.
[30] Maps 4/59, 4/60, 4/63, 4/65.   [31] Maps 5/92, 5/93.
[32] Though the Mileham act affected areas to the south of the Tittleshall division of the Coke estate, it had nothing to do with the former Tittleshall strips. See Castleacre and Tittleshall maps.

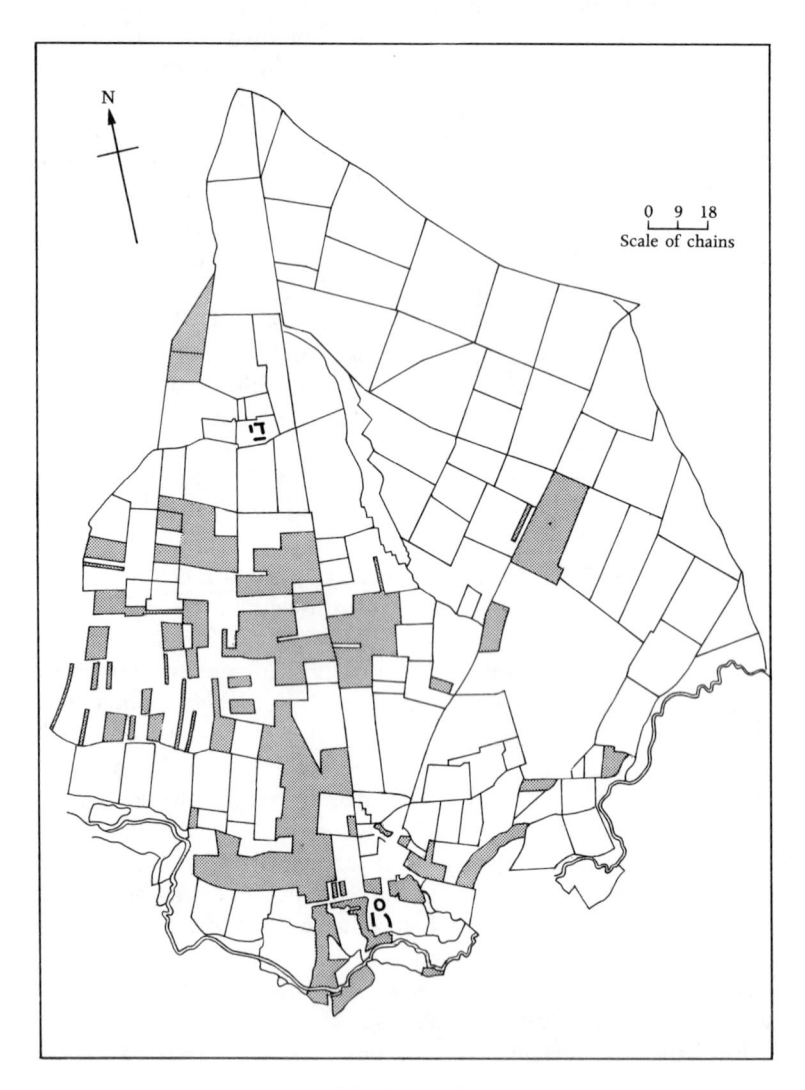

44   *The estates under Thomas Coke, first earl of Leicester*

CASTLEACRE

All the properties not in Coke's ownership in 1779 are shaded. Nearly all had been bought by 1796.

The estates under Thomas Coke, first earl of Leicester 45

purchase of holdings are easy to see on maps, and their effects are not too difficult to estimate. In Castleacre, for instance, Francis Anderson paid about 7s. 6d. an acre in 1714 for his 239 acres of enclosed land, and less than 4s. an acre for his 222 acres of field land.[33] In 1749, also in Castleacre, Thomas Abell paid 7s. 6d. an acre for 523 acres of enclosed land, and only 5s. an acre for 167 acres of open-field land.[34] That complete freedom was not available to holders of open-field land is emphasized by the distinction, drawn in a document of 1751, between the common fields or infield of Massingham and the other arable lands in Carr's farm. The strips were 'to be plowed & sowed according to the Courses & Shifts of the Fields where they lye', the other lands were subjected to much more detailed conditions.[35]

It is much more difficult to see on maps, or to estimate, the effect of the expansion of the cultivated area, or of the more intensive cultivation of formerly lightly cultivated areas. Yet it seems certain that such expansion and intensification were the chief reasons for increasing profitability on the Coke estates in the eighteenth century. Maps by themselves are inadequate evidence. The appearance on them of permanent enclosures where none existed before certainly suggests an increase in the extent or intensity of cultivation. The maps of about 1780 point to such increases. But the existence of what look like enclosures in the outlying areas on early eighteenth-century maps is no proof that these were permanent enclosures or that the relevant land was being as highly cultivated as other land. Other evidence, however, makes it plain that much land treated as second-order land in the early eighteenth century had ceased to be so by the end of the century and that, though the process went on throughout the century, it had had substantial effects in its first fifty years or so. No doubt such land often failed to equal the quality of the best arable, but it came to be capable of producing the same crops in the same types of rotation.

The Abbey farm in Castleacre, contained, in 1714, 947 acres at a rent of £220. Of this farm the Arundell's fold-course made up 475 acres. Every year about 130 acres of the fold-course was under tillage but the rest was left laid down for sheep's pasture. The map

---

[33] Map 5/80. Certainly 52 a. 0 r. 37 p. of his enclosures were pasture and 16 a. 2 r. 23 p. meadow, which reduces the disparity.
[34] Castleacre Deeds, 130.
[35] In Massingham Deeds, 236, 'Draft of Mr. Carr's Covenants for his new lease'.

on which these details are given shows the fold-course divided into breaks. The next map, corrected to 1757, shows a substantial area that had been thus divided as undivided. On the other hand, such divisions as remain show some similarity to earlier divisions. In the map of 1779 the whole fold-course is divided by what appear to be permanent fences.[36] Probably, then, part of the fold-course had been permanently enclosed and brought into cultivation of the same type as that of other arable enclosures by 1757, but the remainder only between 1757 and 1779. It is significant that, while the rent of Arundell's fold-course was £70 in 1714, or just under 3s. an acre, it was £130 in 1755, and that, even by 1727, the rent of the whole farm of which the fold-course was part had risen to £314.[37]

In Tittleshall sometime between 1725 and 1728, Thomas Haylet, junior, held 570 acres, 231 acres of which was 'break'. All of this was counted, together with 'arable enclosures' and 'field lands now partly enclosed', as 'Arable Land'. Of the 'break', 'new closes' were 179 acres. Clearly, land was being improved. In this north-western corner of Tittleshall, the process went on. By 1779 the 'North Lyng' of 60 acres, the 'Steakmoor Thorn' of 13 acres, and the 'Steak Moor Common' of 88 acres had been split into neat enclosures—mostly of about 10 acres. In 1725–8 it had been common or waste.[38] In 1725–8 there were 445 acres common or waste in Tittleshall; by 1779 there were 35 acres.[39] In this parish, two tenants, both named Haylet, held farms of about the same size. One was that just described, held by the younger Haylet, which included land being improved. The other, held by the elder Haylet, was made up of older enclosures. The rent moved in a different way on these two farms. Between 1726 and 1747 the value of the land held by Haylet senior rose by about 2½ per cent, the improvable land of the younger Haylet by over 80 per cent.

In Flitcham, the Abbey farm, around 1723, included a very large area of break and sheep-walk. A valuation of that time includes a great tract of 668 acres, described as 'residue of the break or Sheep's walk', valued at only 3s. an acre. On the other hand, there was already 'break' land valued higher; the break next to the 'Braky field' and the 'Ten Acres' were together 29 acres at 4s. an acre; the

[36] Castleacre maps, 5/80, 5/81, 'Plans'.
[37] A/B 1727, 1755.
[38] Maps 74/4 and 'Plans'. See Tittleshall map.
[39] G.E.D., 77.

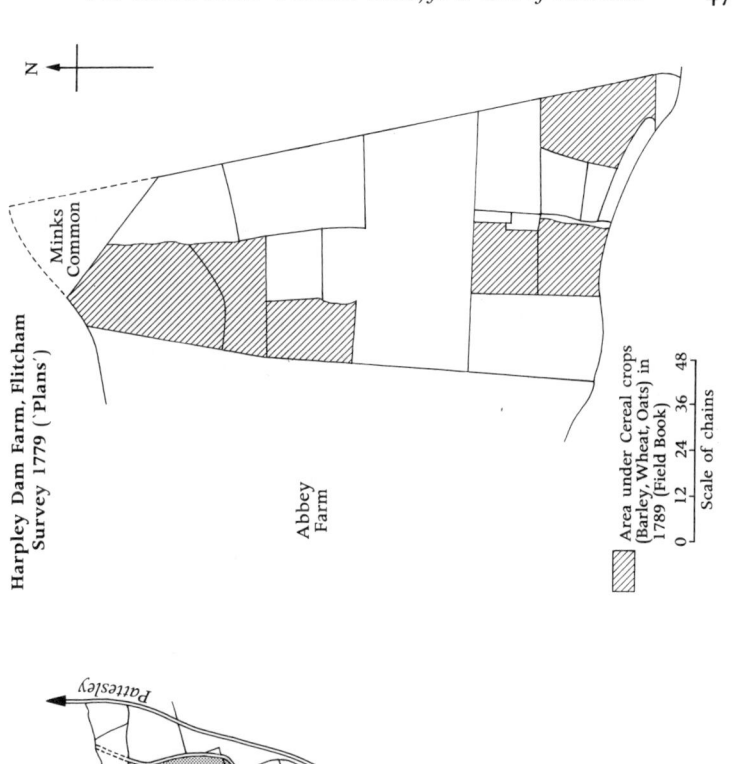

The estates under Thomas Coke, first earl of Leicester 47

Harpley Dam Farm, Flitcham Survey 1779 ('Plans')

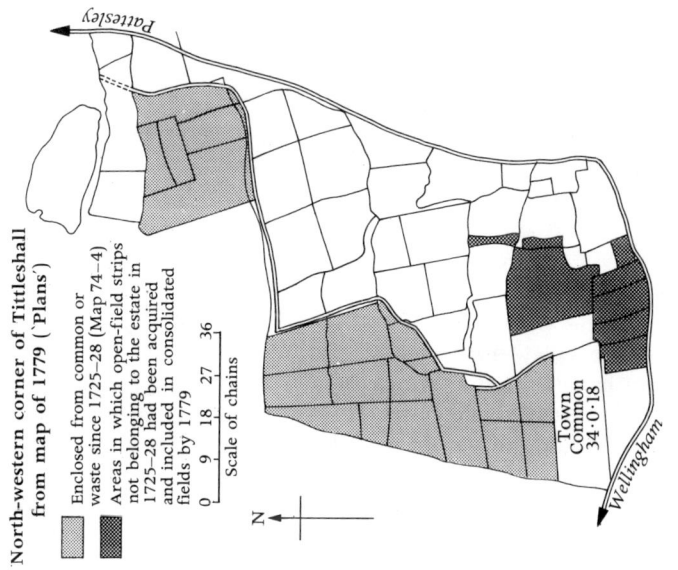

Minks Common

Abbey Farm

Area under Cereal crops (Barley, Wheat, Oats) in 1789 (Field Book)

Scale of chains
0 12 24 36 48

North-western corner of Tittleshall from map of 1779 ('Plans')

Pattesley

Town Common 34·0·18

Wellingham

Enclosed from common or waste since 1725–28 (Map 74–4)

Areas in which open-field strips not belonging to the estate in 1725–28 had been acquired and included in consolidated fields by 1779

Scale of chains
0 9 18 27 36

'Forty-acre break' of 34 acres rated 4s. 6d. an acre and the 'Ten Acre Break' of 8 acres was 6s. an acre. Probably some 'break' had been improved—and there was more to come.[40]

Between 1728 and 1733 there were three substantial farms in Flitcham, Flitcham Hall farm of 260 acres, Little Appleton farm of 200 acres, and the Abbey farm and fold-course of no less than 2,180 acres, including nearly all the inferior land.[41] Up to his death in 1733, John Franklin held the Abbey farm and fold-course together with the Snoring fold-course. He paid £400 a year. On Franklin's death, the farm was taken into hand and the rent charged, for accounting purposes, to Lord Lovell.[42] A readjustment of the farm began; first £128 worth a year was detached from it and let. Then, by 1740, Lord Lovell created an entirely new farm out of the inferior lands of the Abbey farm.

A new barn, new fences, ditches, and hedges were erected. Men were paid to uproot furze and clear the ground to make it ready for ploughing. A new house and stables were built for future tenants, and another house for their shepherds. The land was marled, and was stocked and worked on Lord Lovell's account from 1735 until a tenant took over in 1740. Apart from the temporary investment in stocking the farm, Lord Lovell spent £1,456 on its creation. Its improvement continued by agreement with the tenant. In his first year, 1740, William Dewing was allowed £30 for one year's marling, which was still being allowed in 1745.[43] These lands brought into cultivation in Flitcham were, as nearly always elsewhere, made up of some of the less attractive land in the parish. The 1,000 acres or so of the 'New farm' or, as it soon came to be called, the 'Harpley Dams farm', was a 'large light land occupation'. On the other hand, the late creation of the entire farm meant that all the fields were well laid out, with straight fences and rectangular fields. Even in 1851, though, the farm yielded an average rent of only 12s. 10½d. per acre. As was noted in that year, such naturally unproductive soil could not be farmed at a profit without extensive capital.[44]

Land of this sort could come into full use only after the application of capital and the use of new crops. The capital to improve land came in part from the landlord, in part from the tenant—here per-

---

[40] Norwich P.L., Flitcham Deeds, 472.
[41] Norwich P.L., Flitcham Map, MS. 4295.
[42] Thomas Coke, once again, became Baron Lovell in 1728.
[43] A/B 1732-45. See Plan of Harpley Dam farm.
[44] Keary, i. 129-32.

haps more from the landlord than elsewhere, since it was a question of developing an entirely new farm. As we have seen, there were new crops to exploit. The later burst of expansion of the Napoleonic wars was fostered by rising prices. In the 1730s new techniques and the fall in the return on capital from newly bought land, as a result of the fall in interest rates and the rise in the price of land, made it well worth while for the landlord to apply capital to intensify the use of his land. A rise in the price of land stimulates intensive exploitation of land already owned. It is probably true, in addition, that landlords were compelled in these years to make special efforts to secure good tenants, men with the capital and skill needed to work marginal land in a time of low prices. Certainly there seems to have been some difficulty in letting the new farm: in 1738 Clement Ives paid £10 for being 'released' from Flitcham New farm,[45] and it was not let until 1740. Then William Dewing took it for fourteen years at a rent of £240 a year. In 1755 he took the farm for another twenty-one years at £280 and in 1775 another 21-year lease was granted at £315.[46]

In 1733 total rents from Flitcham had been £626. By 1741, after the partitioning of the Abbey farm and the letting of the new farm, rents were £816.[47] No land was bought in Flitcham in those years. These figures imply a gross return of about 12 per cent on this particular improvement. By 1757 total rents in Flitcham were £1,078. Before then, it is true, some land had been bought, probably worth at least £100 a year.[48] Evidently, intensification of cultivation of former second-grade land was mainly responsible for the rise in Flitcham rents. The New farm, taken from former sheepwalk, was subjected, probably from the first, and certainly from 1774, to the same conditions as other farms.[49] In the conditions of the lease drawn up in 1774, the tenant was to leave, at the end of his term, 100 acres of clover and ryegrass of one year's lying, 100 acres of two years, 100 acres of seeds sown with the last crop of spring corn, and 80 acres of turnips. He was, as was usual then, forbidden to take above three crops of corn and one of turnips

[45] A/B 1738 (accounts current).
[46] A/B 1741, 1755, 1756, 1776, 1777.
[47] A/B 1733, 1741.
[48] A/B 1757. Norwich P.L., Flitcham MS. 489. Tenements possessing common rights were being bought. In consequence 166 acres ceased to be commonable. Cf. 'Plans and Particulars of Norfolk Estates' (1779).
[49] P.R.O., C. 12/572/13, f. 6.

C

before laying down the land for two years, and was not to take above two crops of corn and one of turnips, if the land was laid down for only one year. A distinction was drawn between the 'homegrounds' and the 'marled lands', but both were subjected to the same regulations—the term 'marled lands' implies recent renewed improvement, and the provisions as a whole suggest that the farm was subject to a respectable and productive six-course rotation. It is notable, too, that much enclosure of strip land in the centre of the village took place between 1728 and 1779,[50] by which time Flitcham was almost wholly enclosed in its nineteenth-century shape. The rationalization and improvement of agrarian conditions in Flitcham in the years 1730 to 1779 was dramatic and complete.

Carr of Massingham, in the 1730s held about 800 acres, including a fold-course.[51] The rents rose from £135 in 1708, to £170 from 1724, and to £190 from 1725, which was raised to £280 from 1753.[52] In 1728, when the elder Carr died, much of the farm was already highly cultivated. The inventory of his personal property made on 23 October 1728[53] shows that 6 lasts of wheat (a last is 20 coombs, or 80 bushels), 3 lasts of rye, 2½ lasts of white peas, 3 lasts of grey peas, 30 lasts of barley, and 15 coombs of oats had been harvested. In addition, there were £20 worth of vetches and £30 worth of clover hay. Finally, there were 80 acres of turnips. It seems probable that there was a flock of over 300 sheep.[54] His stock and crops were valued at over £1,250. In 1732 the younger Carr agreed to leave 'all the foldcourse Break or such part thereof Inclosed and shall come in course, well plowed Harrowed and Tilled in the Third Earth' at the end of his tenure. By 1751, when arable lands, enclosures, or breaks were to be sown on a six-course rotation, that group of lands was treated as a unit in contrast to the arable lands in the open field. Carr pointed out then, complaining of the inflexibility of the proposed clauses, that there was 'but one Field uninclosed'.[55] In fact, the landlord was suggesting that all the land should be cultivated on a six-course rotation, fold-course included. Clearly this was something different from fold-course lands in their unimproved state as we have seen them elsewhere

[50] Norwich P.L., Flitcham MS. 4295 and 'Plans' map.
[51] Map 5/85.
[52] A/B 1708-54.
[53] Norwich Castle Muniment Room, Case 33, Shelf F/18, no. 74.
[54] The document is in bad condition. There can be seen '3 hund' at the edge of a disintegrated portion.   [55] Massingham Deeds, 236.

and the money spent on marling and enclosures on the farm of the Carrs shows how it became so, together with rotations including clover and the vital turnip which made it possible to maintain as many sheep as before without devoting great tracts of land wholly or mainly to their nourishment.

Apart from commons, where it might be difficult to secure unrestricted use, most of the second-order land of the type just discussed seems to have been improved by the end of the third quarter of the eighteenth century. An example of legal impediment to improvement is perhaps visible in Massingham. There survived there, surrounded on three sides by land which the Carrs had improved in the first half of the eighteenth century, a common of about 136 acres: 'The Heath' or 'Carr's Common'. This land remained almost useless until it was broken up in 1939.[56]

The *Gentleman's Magazine* for October 1752 contains an important contribution describing the improvements made in Norfolk, more particularly in the 'western parts'. The writer stresses the value of marling, of artificial grass and clover, of turnips, and of enclosures. He links the use of marling, and the sowing of turnips and clover, with a lowering of the price of wool, and an increase in the price of corn for export, as reasons for the start of the improvements he describes. The reduction of the price of sheep made it desirable to increase the amount of land devoted to arable cultivation. Norfolk, he points out, was particularly well situated for exporting grain to Holland, where, he argues, the importation of corn from Danzig had declined, partly as a result of devastation caused by wars in the Baltic and Poland, partly as a result of a greater demand for corn in Sweden after the Russians had taken the Swedes' corn-growing regions from them and limited exports therefrom.

According to this writer, the improvements had been gaining ground for about a hundred years. (It is possible, and the static figures for Coke rents in the seventeenth century support this speculation, that the Coke estate was catching up after 1707 on other estates in Norfolk.) On open-field farms, he goes on, it was impossible to benefit from clover and turnips, though some open-field farmers had marled their lands. Not all farmers of enclosed

[56] See Major Keith's account in *Journal of the Royal Agricultural Society of England*, 103 (1942), 3-4, and his illustrations which show land doubtless similar to that tackled by eighteenth-century improvers.

lands made adequate use of clover or marl, but nearly all had taken to turnips. Some of the improvement had taken place on land enclosed from open-field, but 'more commonly it is upon our break lands we improve'. This land had formerly been used for sheep's feed alone for 7, 10, or 15 years. It would then be broken up, yielding a crop of rye followed by oats or barley, and the land then allowed to revert to sheep-walk while other such land was broken up in its turn. This kind of land was first improved by marling, for which pits could usually be dug every 30 or 40 acres, pits which subsequently became ponds, one for each enclosure, of great use when cattle were fed in the enclosures. The new hedges provided shelter. On newly improved land turnips were grown first, which, duly hoed, cleaned the land and mixed the marl with the old surface. The turnips were fed off on the soil. The next crop was barley or oats with which clover seeds were sown, the produce of which was subsequently pastured or mown for hay. The next year wheat followed on the broken-up ley and, if the crop was clean, a barley crop would follow, or, if not, turnips again. The writer gave some sample rotations on newly improved land. These mostly include two-year clover leys.[57]

Evidence of price movements in Norfolk is scanty. Such evidence as there is at Holkham does not suggest that prices for corn were rising significantly between the early 1730s and the middle of the 1750s, but the pattern of demand suggested by this writer may have prevented falling prices suffered in other parts of the country.[58]

Philip Miller, the author of a popular *Gardeners Dictionary*, offers similar evidence. In the 1733 edition, he wrote of turnips that their use for fodder in winter 'is become a great improvement to barren, sandy lands, particularly in Norfolk, where by the culture of turnips, many persons have doubled the yearly value of their ground'. In the 1759 edition was added that 'this plant was not much cultivated in the fields till of late years, nor is the true method of cultivating turnips yet known, or at least not practised in some of the distant counties of England at this time,' and that

... in Norfolk and some other counties they cultivate great quantities of turnips for feeding of black cattle, which turn to great advantage to their farms for hereby they procure a good dressing for their land, so

[57] *Gentleman's Magazine* (1752), 453-5; rotations are in the November issue, p. 501.
[58] Holkham Deeds, 1067 (Hall farm accounts); 'Posting book for the Years 1749-54' (Muniment Room).

that they have extraordinary good crops of barley upon their lands, which would not have been worth the ploughing, if it had not been thus husbanded.

The author of the *Complete Farmer* (1767) substituted for Miller's phrase 'till of late years' the words 'till within the last sixty or seventy years'. Miller's eighth edition, of 1768, however, implied an earlier start for the field use of the turnip: 'this plant was not much cultivated in the fields till within a century past.'[59]

This literary evidence supports the argument that there were important changes in productivity, particularly linked with the use of turnips as a field crop, which can be ascribed to the period 1670 or 1680 to 1760, and that these changes were particularly significant and especially early in north-western Norfolk.

iii. *Investment in, and administration of, the estates*

Large farms were familiar on the Coke estates in the first half of the eighteenth century. Some were very large indeed. Edmond Skipton, for instance, held 1,000 acres in Fulmodestone at £300 a year in 1707.[60] But often much of the land that made up these large farms was poor quality land, used wholly or partly for sheep-walk alone. Before 1725 Thomas Pigg held 735 acres in Longham, but 209 acres were 'fold course or heath', as they remained until the Napoleonic war.[61] The size in acres of a farm is not in itself a reasonable measure of its economic value and a 'large farmer' must be a man holding a valuable farm if a differentiation between 'large' and 'small' farmers is to have any useful significance. In 1727 there were twenty-six farms in Norfolk, owned by Coke, which were let at a rent of £100 a year or more, of which four were above £200 a year but below £300 and two above £300. By 1758 there were forty-six farms let for £100 or more, of which thirteen were above £200 but below £300 and one above £300. The area of farms on the estates sometimes grew less as time went on as marginal land was improved, cut off, and farmed separately; the value of farms tended to increase.

The farms on the estate were for the most part on lease in the

[59] Philip Miller, *The Gardeners Dictionary*, 2nd edn. (London, 1733), 7th edn. (1759), 8th edn. (1768), s.v. 'Rapa'. Anon. [A Society of Gentlemen], *The Complete Farmer* (London, 1767), s.v. 'Turnep'.
[60] A/B 1707.
[61] Map 93/5 and 'Plans'.

first Lord Leicester's time. The usual term was 21 years, long enough for tenants to be encouraged to improvement without the landlord losing control of his farms and depriving himself unduly of opportunities to share in increasing profits: 'without leases', wrote Kent later, 'no marling, to any extent, would have been undertaken, nor so much ground brought into cultivation by one-third as there now is.'[62] In 1727 there were twenty-five 21-year leases in force together with three for shorter terms. In 1751 there were thirty-two leases for 21 years and two for shorter terms.[63] It is not certain when 21-year leases became usual. The granting of leases was evidently normal under the guardianship,[64] though the guardians were legally unable to grant leases running beyond 1718 when Thomas Coke would come of age. A document refers to leases made by the guardians to Norfolk tenants in 1715 for 3 years and 1716 for 2 years—they would expire in 1718 when Thomas would be 21.[65] And in March 1714 a farm was let for 14 years 'vizt. 4 yeares from the Guardians—& Mr. Smith is to use his indeavour with Mr. Coke when at age to confirme the lease for the other ten years'.[66]

The terms of leases are of great importance. They provide evidence of the standards of cultivation that the landlord was setting to his tenants. The progress, and the technical changes, in the clauses dictating how the land was to be used, are evidence for the progress of farming. Certainly it is difficult to say how far such clauses did or did not call for farming practices different from those that the tenants would in any case have employed. It is also uncertain how far their observance was rigidly enforced, although Carr in 1752 took the clauses proposed for his new lease seriously enough to list his objections to them.[67] At the lowest estimate they provide a summary of the levels of technique assumed by the landlord. One may reasonably believe that they were one of the most important means by which landlords influenced the conduct of their tenants and advanced good husbandry. John Mordant's remark[68] that stewards should see that 'the tenants keep their different farms up in a due course of good husbandry ... should see that all tenants

---

[62] N. Kent, *General View of the Agriculture of the County of Norfolk* (1796), 123. Leases were perhaps more important for marling when he wrote than earlier; for its cost was by then borne by the tenant.
[63] A/B 1727, 1751.
[64] See e.g. Holkham Lib. MS. 743, p. 78.
[65] Holkham Deeds, 1052.
[66] Holkham Lib. MS. 743, pp. 156–7.
[67] In Massingham Deeds, 236.
[68] *The Complete Steward* (London, 1761), i. 381–2.

The estates under Thomas Coke, first earl of Leicester

punctually perform all and every covenant which they agreed to in their leases' perhaps shows that Mordant thought it possible to enforce such clauses—to make them more than formal admonitions. The evidence of what was contained in leases of Lord Leicester's time is unhappily scanty.

An agreement of 1721 for a 21-year lease called rather tentatively for the use of artificial grasses and short leys: the tenant agreed not to 'take above three Crops of Corn or Grain before the same be laid down with Grass or till'd for Summerland. And to sow with Clover Rye Grass or Nonsuch all such lands as are laid down with Grass and not to plow up any such lands till the same have so lain for two years.[69] Even earlier, in 1696, an agreement for a lease of the farm in Waterden had laid down that the tenant was 'not to plow or sow any of the Arrble. land out of Course; nor to plow or Sow any of ye Infield Land more than five Cropps and the Breakes but three only, before ye same shall be somertilled or laid for pasture': a clause similar to those of the twenties and thirties, though notably less strict.[70] Again, in 1714 a tenant was restricted to four crops from infield and three crops from sheep-walk before laying down to grass.[71] In 1732 the 21-year lease granted to Carr of Massingham provided that the tenant should not 'sow above three Crops of Corn successively before the land is Sumertilled & sown Turnips or laid down with Grass to be fed'[72] and a lease agreement made in 1733 insists that the tenant 'shall at no time take above four Crops without laying the ground down with proper Grasses for two Years'.[73] In 1751 the draft clauses for Carr's new lease included the provision that the tenant should not sow 'above three Crops together and one of those after Turnips or Summertillings before [the land is] laid down with Grass seeds to lye two Years'.[74] This implies a six-course rotation: corn, corn, turnips, corn, grass, grass—a perfectly reputable rotation by the standards of the end of the eighteenth century, and one which it is striking to find demanded so early. Such newly introduced clauses were obviously more than mere time-honoured conventions in the drafting of leases and at the least, they formally proclaimed high standards of cultivation.

[69] Tittleshall Deeds, 131. This is the earliest specific reference to the use of ryegrass.
[70] Waterden Deeds, 58a.   [71] Holkham Lib. MS. 743, pp. 156-7.
[72] In Massingham Deeds, 236.
[73] Holkham Deeds, 1072.
[74] In Massingham Deeds, 236.

A large amount of money continued to be laid out on the estates. In the years 1722 to 1759, £49,809 was accounted for as spent on repairs and improvements or allowed to the tenants in Norfolk. This was just over 17 per cent of gross rents in those years in Norfolk. The proportion of gross rent that was thus returned rose to a peak in 1735-9 when the proportion of gross Norfolk rents returned in repairs and improvements was no less than 27 per cent. In 1740-4 it was 16 per cent; in 1745-9 it rose again to 20 per cent but in 1750-4 it was down to 13 per cent and in the last five years of Lord Leicester's life only 11 per cent was spent thus. Cauldwell claimed that he reduced the outgoings on repairs and improvements during his stewardship (which began in 1742) by shifting some of these burdens on to the tenants.[75] This may explain the fall in investment. It might be tempting to imagine that the heavy expense on building the new house at Holkham led to skimping on the estate but the evidence suggests that this is unlikely. Probably less investment was made because less was needed. Later on, as we shall see, the amount spent on repairs by tenants and recorded in the Audit Books was supplemented by moneys paid out from the Holkham office or by materials provided by it. It is impossible to say whether this was going on in the years before 1759. If it was, the figures for repair spending above would understate the actual amount laid out.

During the period 1722-59, spending on 'improvements', just over £24,000, was nearly equal to 'repairs', about £25,700. If the spending on 'improvements' can be regarded as generating the increase in rent in these years attributable to causes other than new purchases—an increase of the order of roughly £2,400 or so—then, very tentatively, a return on this investment of about 10 per cent can be suggested. Such figures must be used with great caution, since they raise various problems. Though the early eighteenth-century accountants on the Coke estates (unlike those after 1784 when the word 'improvements' disappears) distinguished between 'repairs' and 'improvements', it is not certain how carefully and consistently a line was drawn between them. Thus in the Holkham 'improvements' for 1745 one item includes payments for 'making new banks, mending old banks'. Again, payments were included under the heading of 'improvements' which were not capital investment at all, such as payments to outgoing tenants for crops left

[75] P.R.O., C. 12/572/13, R. Cauldwell's 'Answer'.

growing on their farms. Many payments were made for marling, which may or may not reasonably be thought to fall under the heading of fixed investment. If payment for marling was made the responsibility of the tenants, with corresponding reduction in the rents (a change which in fact took place on the estate in the later eighteenth century) then the process would seem, in the accounts, to involve less investment on the landlord's part, though the economic facts might be unaltered (although when the responsibility for finance lay on the landlord it would mean he had more of the initiative in deciding when, where, and how to invest).

In spite of the reservations that should be made, it is clear that the great bulk of payments for 'improvements' were on account of fixed capital in the strict sense: for new building, ditching, and fencing. It is certain, therefore, that Lord Leicester and his servants took an active and perhaps preponderant part, financially and administratively, in the improving of the land in the hands of the tenants.

iv. *The home farm*

Soon after Thomas Coke came into control of his estates, a home farm at Holkham began to be a regular institution. During the guardianship, Humphrey Smith, the chief steward, held what had perhaps earlier been a home farm, the 'Hall farm', at a rent of £75 a year. Together with this, he had held other lands bringing his total rent to £264. 11s. 8d. a year.[76] These lands were taken over as a farm in hand in 1722. The farm was at once stocked with 2,232 sheep (542 ewes and 1,690 wethers).[77] Rye, wheat, barley, turnips, nonsuch, lucerne, and clover were grown on it from the earliest years.[78]

In 1724–8, the first five years of the home farm (allowing one year for settling down), farming operations made an average yearly profit of £389. In 1729–33 the average profit was very slightly less, £387 a year. In the next five years, 1734–8, it went up to £610 a

[76] A/B 1707–17.
[77] Holkham Lib. MS. 740. Disbursements in Norfolk, 1722–3.
[78] 'Geo. Appleyard's Accot. of the Domestick Disbursements in Norfolk' (Game Larder) for the years 1724–8 has payments in 1724 for rye seed, wheat seed, turnip seed, ryegrass, nonsuch, and clover seed and much barley was sold then. J. E. T. Rogers, *A History of Agriculture and Prices in England*, vii, 1703–93, Pt. 2 (Oxford, 1902), 636–704, printed the accounts for the Hall farm from Michaelmas 1731 to Lady Day 1737.

year. At Michaelmas 1737 a substantial area was apparently carved out of the home farm: over 850 acres, which formed the new Branthill farm, let at £236. 10s. 0d. a year. The reduced Hall farm produced a profit of about £450 a year in the next five years, 1739-43. In 1744-8 profits rose to about £575 a year and in 1749-53 to £760. In the last five complete years of Lord Leicester's life annual profits were just over £900. The area of the Hall farm changed from time to time, and this explains in part the striking increase in the return from the farm. In 1746 the farm contained just over 1,100 acres. Of this, 284 acres were wood and water, 150 acres new or old marsh, about 100 acres 'common lings and whins', and 594 acres was arable land in the park. By 1759 there were over 200 acres more of arable land taken away from adjoining farms.[79]

The economy of the Hall farm can be analysed for the early fifties.[80] The figures which emerge are recorded in Table B.

TABLE B

Receipts	1750			1751			1752			1753		
	£	s.	d.	£	s.	d.	£	s.	d.	£	s.	d.
Arrears of corn	192	5	9	96	7	3	186	0	6	242	18	7
Wheat	87	10	0	33	13	9						
Rye and oats	34	14	11	23	15	5	332	0	8	460	9	11
Barley and malt	328	18	4	391	14	9						
Bullocks and dairy	314	6	0	252	3	0	328	18	11	303	4	7
Sheep and lambs	173	14	7	158	9	9	164	17	4	185	19	4
							Hogs and horses					
Hogs	68	9	9	51	9	5	29	7	6	40	0	6
							Carriage and limekiln					
Horses and limekiln	281	11	7	403	7	5	241	16	4	368	5	7

A substantial proportion of these moneys credited came from the Coke household; it was credited to the farm account and debited to others, such as the household account or the building account. Thus the item 'horses and limekiln' included receipts for carriage done or material supplied for the new building and substantial quantities of barley and malt were sold to Lady Leicester. Hay sold was included, mysteriously, under barley and malt. In 1751 £67. 10s. 0d. was credited for hay, most of which went to the Holkham stables. The figures show that barley was by far the most

[79] A/B 1724-59.   [80] 'Posting book' for the years 1749-54 (Muniment Room).

important crop grown. The receipts for bullocks and dairy reflect the acres of marshland grazing exploited with the farm and they are not typical of farms in north-western Norfolk. (Of 1,164 sheep sheared in 1732, 345 were maintained in the Holkham marshes.)[81] Payments for the same years were as recorded in Table C.

The rent paid was credited to the estate rental account, and in the annual audit books, the amount due from the Hall farm was noted as if it were a farm in the hands of an ordinary tenant. The figures given earlier for the profits of the farm exclude from the debit side the amount 'paid' as rent. A high proportion of those profits was credited to the rent account. The nominal rent charged tended to increase as time went on. In the 1720s it was about £250 per annum; in the 1730s more than £300, until Branthill farm was taken away, when the rent fell to £102. This was raised to about £250 in 1740, which rose to above £400 by 1749. In the 1750s the rent was about £420 a year.

The rent charged for the farm gives some idea of the value attached to various kinds of soil: in 1746 'Wood and Water' were valued at 5s. an acre; the main bulk, the arable land within the park, at 7s. an acre; 'common lings and whins' at 2s. 6d. an acre; old marsh at 10s.; new marshes and some other old marsh at 5s. per acre. The rise in the rent charged in the forties and fifties was caused

TABLE C

Disbursements	1750	1751	1752	1753
	£ s. d.	£ s. d.	£ s. d.	£ s. d.
Labourers' weekly bills	49 7 0	62 5 9	82 13 4	71 3 8
Labourers' other work	87 11 4	92 3 5	95 18 9	95 13 10
Expense of seeds bought	111 5 7	45 16 4 }	296 13 9	300 1 2
Expense of stock bought	31 10 0	124 12 0 }		
Tradesmen's, etc., bills paid	46 17 5	64 6 11	51 2 9	53 11 5
Limekiln and malthouse	23 7 0	100 13 2	53 4 5	74 5 4
Rent, tithes, and taxes excl. wood and water	375 9 8	370 13 5 }	484 14 1	519 7 1
Servants' wages	158 3 5	177 17 1 }		

[81] Holkham Deeds, 1067.

by more land being added to the farm, not by a change in the value attached to the land. Some of the arable land taken over from adjoining farms was accounted for at 7s. an acre, some at 8s. 3d.; the main block of arable land continued to have assigned to it the value of 7s. per acre.

The analysis of disbursements shows how high a proportion labour costs bore to total farming costs. Unfortunately one cannot be certain that in its management and financing the Hall farm was characteristic of farms in north-western Norfolk—one may reasonably suspect that it was. However, the fact that many of its payments were made to other departments of Coke administration and that many of its receipts came from the Coke household must have involved some divergence from the sort of management one would have found in a self-contained farming enterprise.

The rent may or may not have been what a competent and thrifty tenant would have paid, though very likely it was an honest estimate of it. The rent charged reduced the theoretical gross profits given above to a theoretical net profit of about £195 in the 1730s, £209 in the 1740s, and £440 in the 1750s. In some ways these figures should not be taken very seriously. The internal payments were notional ones; if—as an instance of their possible unreliability— a steward wished to suggest that the farm was doing well, he could charge the stables and the kitchen unduly large sums for what they took. Again no account was taken of the value of the stock held at the beginning and end of the periods covered by the accounts. Still, they give some idea of the relation between outlays of various kinds and the receipts for different commodities and an impression of the order of magnitude of farming profit in the first half of the eighteenth century.

The farm was a serious and viable enterprise in the first half of the eighteenth century. Its existence alone is sufficient to destroy the old picture of the uncultivated, utterly barren character of this part of Norfolk before the time of Coke of Holkham. In general, indeed, examination of Lord Leicester's activities as a landlord makes it clear that much of the so-called agricultural revolution took place on those estates during his period of tenure. It is a matter of intellectual rather than agricultural history that makes Coke of Norfolk famous as a landlord while Lord Leicester is remembered only as a virtuoso. One aspect of the romantic movement was that farming became fashionable; Lord Leicester's estates were for profit not for show.

5

The Estate and Finances under Margaret, countess of Leicester, and Wenman Coke, 1759-1776

LORD LEICESTER died on 29 April 1759. He had signed his will in May 1756 and added a codicil in May 1757. His will was proved on 2 June 1759.[1] He gave all his real estate, including leaseholds and the lighthouse, to Sir Matthew Lamb and Ralph Cauldwell as trustees. The codicil brought in estates purchased between May 1756 and May 1757. The trustees, firstly, were directed to raise money to pay off the debts not on mortgage: hence the mortgage of 20 May 1765 for £30,616. 18s. 10½d. lent by the Hon. Charles Yorke, Charles Cocks, Philip Pyndar, and William Pyndar (the last two being London merchants).

The descent of the house and estates was laid down by the will, and the estates were held under it for the remaining years covered in this study. Immediately, Lord Leicester's wife, Margaret, was to have the estates for life. After her death, they would go to Anne Roberts, his sister, who had contracted a runaway marriage with Philip Roberts, an officer in the Blues, in 1716. It was emphasized in the will that the profits of the estate were to go into *her* hands. After her, the estates were to go to Lord Leicester's nephew, Anne's son, Wenman Coke (he had changed the name Roberts to that of Coke). Anne died before she could benefit from her brother's will, but her son succeeded to the estates when Lady Leicester died in 1775. Wenman Coke held the estates for only one year. He died in 1776, and his son, Thomas William Coke, succeeded him, as his great uncle's will had directed. Since then Wenman's direct descendants have lived at Holkham.

These descendants were to have to face the great burden of debt left by Lord Leicester; they had to pay for the house they lived in

[1] Office Copy H.F.D., 56B. The Principal Probate Registry reference to this document is P.C.C. Arran 208.

and for the speculations of 1720. Lord Leicester provided that the trustees were to set aside the sum of £3,000 clear of all deductions each year in order to pay off the debts he had contracted before he died, and to continue such payment until the estates were freed from debt. These debts amounted to over £90,000 and the interest on them was therefore at least £3,600 a year, so that, to begin with, over £6,600 would have to be paid yearly for interest and debt repayment. Lord Leicester's personal estate amounted to only £13,440. 4s. 7d., including £5,427. 12s. 4d. from the last half-year's rents and £3,527. 10s. 0d. received for the sale of his house near Berkeley Square. He had a credit balance with Child's of £627. 13s. 11d. and £636. 16s. 11d. with Sir Matthew Lamb, but he owned no securities of any kind and no one owed him any money. After paying legacies, interest due on debt, and funeral expenses, there was apparently only about £200 to use in discharging debts.[2] Lord Leicester, however, exempted his wife from the payment of the £3,000 a year and this sinking fund was not to operate until after her death, but she was, on the other hand, subjected to the obligation to pay £2,000 a year until the buildings at Holkham were complete.

The will gave power for the granting of leases of any of the property (except the house at Holkham) for any term not exceeding twenty-one years in possession with the best improved yearly rent reserved. But on the Somerset estate, leases of cottages and lands of up to £10 annual rental value could be made for three lives or for any period determinable on three lives with only the customary yearly rent reserved (and fines taken at the beginning of the term).[3] The trustees were given power to settle lands to secure a £1,500 jointure for anyone a beneficiary should marry, but they could not mortgage any of the estates. The two trustees were given an annuity of £200 each to compensate for their trouble.

Finally, Ralph Cauldwell was to be kept as steward for the rest of his life at his usual salary, in addition to his £200 as trustee. His salary had been £200 a year since he became steward in 1742, the same salary as that which had been paid to his predecessor, George Appleyard. Evidently Cauldwell had been trusted and liked by

[2] H.F.D., 62.
[3] The limitation to £10 annual value is interesting: it had not appeared earlier. In the deed of 30 Jan. 1721 conveying the western manors to trustees to secure mortgages (H.F.D., 44(i)) the trustees were empowered to give such leases without limit of annual value.

Lord Leicester and the widow completely agreed with her late husband, so that during this period the steward attained a greater position than any other on the estate until the masterful Blaikie took control in 1816. Lady Leicester raised Cauldwell's salary to £300 a year.[4] When Lamb died in 1768, Cauldwell became sole trustee.[5] Lady Leicester's will, drawn up in 1766, included a legacy of £2,000 to Cauldwell, and, bequeathing certain leasehold estates, which were within her testamentary power, to Thomas William Coke (by-passing Wenman), she directed that Cauldwell's £200 a year as trustee should be paid from the rents from them whether he remained steward or not, and if this money was not duly paid, that Cauldwell should become entitled to the estates themselves. In a codicil of 1773, she made Cauldwell joint executor of her will together with her sister, Lady Gower, and bequeathed another £300 to Cauldwell for his trouble as executor. By that time it was probably clear to her that the future occupants of Holkham did not share her partiality towards her late husband's steward, but that would not have lessened her affection for him. In her will, Lady Leicester declared austerely that Wenman Coke's conduct towards her had not been such as she had just reason to expect, and she explained that if he had treated her better, he would have had some of the benefits she now bequeathed to her sister. It seems, for instance, that Lady Leicester repaid some of Lord Leicester's mortgages—presumably in part from the proceeds of the estates— but instead of relieving the estates of them, she gave the mortgages when she died to her sister, thus keeping intact the burdens that Wenman and his son would have to face.[6] It was perhaps fortunate for Wenman Coke and his son that Lady Leicester was not in a position to disinherit them altogether. A codicil to her will incorporated a provision to protect Cauldwell against attack, laying down that if difficulties were made over his executorship and obstacles placed in the way of his receiving his £2,000 legacy, Cauldwell was to become entitled to the furniture and goods she had bought at her own expense for Holkham.[7] She left a list of this furniture bought by her, between 1759 and 1769, for a total of £3,096. 5s. 8d.[8]

[4] A/B 1722-59, 1775.
[5] P.R.O., C. 12/572/13, bill.
[6] G.E.D., 78.
[7] H.F.D., 58. Office copy of Lady Leicester's will, dated 15 Sept. 1766, and codicil, dated 20 July 1773.
[8] In Library, account book: 'Lady Leicester's Jewels and Furniture.'

Wenman Coke and Thomas William Coke may well have felt that the old lady and Cauldwell had been leagued together against their interests. Cauldwell's entry of the payment to himself of his salary in the only audit account of the Grand Estate presented to Wenman Coke has a defensive tone: his £300 is declared to be 'as in his former accounts allowed by Lady Leicester', but Wenman is a nebulous and ephemeral figure in the history of the estates and little is known about his relations with Cauldwell. It is certain, though, that his son, Coke of Norfolk, turned on Cauldwell soon after he came into the estates. Though Cauldwell gave himself his £300 salary, 'as allowed in former Accots.', in 1776 and may have got it through, by 1777 he was back on his old salary of £200 a year.[9] The attack on Cauldwell was doubtless stimulated by a newly appointed servant of Thomas William Coke, who rose to a brief eminence in 1776. On 1 August 1776 Richard Gardiner was appointed to the previously unheard-of post of Auditor-General over all Coke's estates in Norfolk.[10] Gardiner had clashed with Cauldwell before then: in 1772 he had vigorously pressed a request for the lease of the farm in Massingham at the expiry of the lease to the younger Carr, and had secured Thomas William Coke's promise of the farm if he should be in control when the farm became vacant. In case he was not, Gardiner secured the support of Nicholas Styleman of Snettisham, whose advice Lady Leicester had promised to take, so Gardiner alleged. Gardiner asked for the farm in May 1772, urging that the air of Massingham agreed with him, Styleman wrote to Lady Leicester in July 1772 to back him up, and Gardiner wrote again himself in November. All failed and Gardiner did not get the farm, which was let to William Blyth.[11] Probably the attack on Cauldwell began before Gardiner's own quarrel with Coke and his dismissal in July 1777.

Cauldwell was attacked on two grounds. Coke claimed that his accounts as executor to Lord Leicester had not been submitted, and alleged that some agreements for hiring of farms he had made in Lady Leicester's time provided for grossly inadequate rents. The estate was thus being defrauded of great sums of annual revenue for rents. It is of some importance to determine whether or not Cauldwell's rents were honest. If he was justly accused of fraud in

[9] A/B 1776-7.
[10] R. Gardiner, *A letter to Sir Harbord Harbord, Bart* ... (London, 1778).
[11] Massingham Deeds, 236.

fixing tenant's rents, the rise in rents that took place under his stewardship would be smaller than it should have been. This would mean that movements in rent would understate the rise in productivity of the lands owned by the Cokes during the first three-quarters of the eighteenth century.

Coke alleged that, in nine contracts for leases of farms made in Lady Leicester's time, Cauldwell had fixed rents of no less than £3,500 per annum below what those farms taken together should have brought in. For these nine farms, Cauldwell had secured rent increases of about £1,000 a year in the new lettings; Coke argued that he should have secured about £4,500. Coke gave as a particular example the large farm let to Benoni Mallett in Dunton, at £680 a year for 21 years from Michaelmas 1769. All the nine farms had been looked over, but Mallett's farm had been given special attention 'by two of the most experienced Surveyors in the Kingdom' who had, it was pointed out to counsel, 'valued Mallett's farm at £1,209. 7. 0. a Year and over and they can let it to indisputable tenants at that sum'.[12] Here is a means for testing Coke's allegations. The rent fixed by Cauldwell was not modified until the twenty-one years ran out in 1790. From then the rent became £1,000 a year. Clearly if that was a reasonable rent in 1790 after a long period in which rents had been rising, £1,200 cannot have been reasonable in 1769. This speaks for Cauldwell. So also does the fact that Coke gave way. In 1779 he agreed to waive all objections to the contracts for leases.[13] Possibly, on the other hand, Coke gave in only to hasten Cauldwell's departure from the stewardship and trusteeship of the estates. For Coke may have discovered that no process could get Cauldwell out without Cauldwell's consent, a consent Cauldwell, in fear of actions against him from those tenants whose agreements were in jeopardy, declined to give without full indemnification for the leases in addition to acceptance of his accounts as executor. After all, Cauldwell was entitled to the stewardship and trusteeship under the same document, Lord Leicester's will, by which Coke was entitled to the profits of the estate. Still it does seem unlikely that Coke would have climbed down over the leases if he had continued to be convinced of the soundness of his argument. The case against Cauldwell we must leave as, at least, not proved.

[12] 'Copy of Case & Mr. Kenyon's opinion relative to Mr. Coke's leases.' In H.F.D., Bundle 18.   [13] P.R.O., C. 12/1073/3.

Cauldwell was disposed of by the end of 1782. Having accepted Cauldwell's lease agreements, Coke appointed an auditor to inspect Cauldwell's estate accounts for 1775 and 1776, which were allowed. In 1780 a Master-in-Chancery was directed to examine Cauldwell's accounts as executor and to arrange for the appointment of new trustees: James Dutton, Coke's brother-in-law (later Lord Sherborne) and Thomas Master. The Chancery Master reported in December 1781 and certified Cauldwell's accounts as correct. Then, in February 1782, the estate was conveyed to the new trustees.[14] Cauldwell was pensioned off with £400 a year, the equivalent of his allowance as trustee together with his former salary, and he received this until his death in 1791.[15]

His reputation seems to have recovered at Holkham; in 1831 Francis Blaikie wrote of 'Ralph Cauldwell, the celebrated Holkham steward' and referred to his 'able management' of the estate.[16] He appears to have had some notoriety in his own time. When François de la Rochefoucauld passed through Hillborough in 1784, he noted that one of the houses belonged to 'a business man who may be regarded as more than usually clever. In the course of a few years he contrived, while administering the property of Mr. Coke's uncle, to put together nearly £40,000, and his skill was such that Mr. Coke was obliged, when he took over his uncle's property, to pay him a sum of money to get rid of him.'[17] Cauldwell held at least one other appointment, quite separately from his work on the Coke estates: this was acting as receiver of the rents of the Fitzwilliam estate in Norfolk, at a salary of £70 a year.[18] This, and his lending to his employer, are in contrast to the exclusive and bureaucratic nature of Francis Blaikie's stewardship after 1816.

We have seen that Cauldwell could claim to have raised the rents in Norfolk by 19 per cent during the time, 1742-59, that he was steward under Lord Leicester, in addition to making advantageous

---

[14] P.R.O., C. 12/1073/3.
[15] A/B 1781. Coke had evidently involved Cauldwell in some loss of income apart from the cut in salary. At the end of 1781 Cauldwell paid to himself £63. 0s. 0d. 'what he paid Saml. Brougham as his agent residing at Holkham to transact the necessary business of the Trust Estates in Norfolk, two years ended Christmas 1781 at £31. 10s. 0d. a Year, for which Mr. Cauldwell was always reimbursed with profit by Fees for leases and of Courts up to Christ^s. 1779, since then those Stewards Perquisites have been withheld and otherways disposed of by Thomas William Coke Esq.' A/B 1782.
[16] A.L.B., i. 145.
[17] J. Marchand (ed.), *A Frenchman in England* (Cambridge, 1933), 213.
[18] Northamptonshire Record Office, Fitzwilliam (Milton) Estate Accounts. F(M), vol. ii.

purchases. The rise in rents continued after Lord Leicester died. In 1758 gross Norfolk rents were £11,027 and when Thomas William Coke took over the estates in 1776 they were £12,332— the increase is about 12 per cent.[19] This is a smaller rise in rents than in the earlier period of Cauldwell's stewardship. Perhaps many of the gains of new techniques had been realized in the earlier period—the movement of rents certainly suggests that agricultural progress in Lord Leicester's time was more rapid, taking price movements into account, than at any other time in the years under study. Purchases of land evidently came to an end when Lord Leicester died—perhaps the widow did not wish to acquire land for the benefit of Wenman Coke and his son—and in that way Lady Leicester's actions as controller of the estates form an exception to an otherwise steadily maintained policy. About other forms of investment it is not possible to speak, since there is a gap in the surviving series of Audit Books for the years 1759 to 1775—these books may have disappeared at the time of the Chancery case. Cauldwell boasted to Chancery that he had shifted some of the burdens of maintenance and improvement on to the tenants, and such action might account for the relatively small rise in rents under Lady Leicester.[20] The agreements for leases made by Cauldwell in the interregnum were listed by him in a schedule to his 'Answer' in Chancery. Tenants agreed to keep in repair houses, wells, and buildings, to pay all workmen's wages, to pay for nails, iron work, materials for thatching and daubing, and for carriage of materials. They were also to repair gates, fences, and hedges. They were to be allowed bricks, paving stones, tiles, lime, hair, and rough timber, and were entitled to cut thorns for wood fencing. Workmen's wages for repairing accidental damage were also to be allowed by the landlord.

The leases, too, included extended provisions for securing good husbandry on the farms. Cauldwell declared that in the making of these contracts

... this Defendant had regard not only to the Advancement of the Rents in present but to the future Cultivation and Improvement of the several farms ... he apprehends and believes that by means of the Terms relative

[19] A/B 1758, 1776.
[20] P.R.O., C. 12/572/13. R. Cauldwell's 'Answer'. In fact there was a rise in average net income before tax in the years 1776-9 of 32 per cent over the years 1755-8, but net income was liable to be erratic from year to year.

to the cultivation thereof, the said Estates are improved in their Value and will at the Expiration of such leases, if the present prices of the produce of the lands shall continue, admit of still further advancement,

a statement which effectively summarized Coke leasing policy before and after Cauldwell's time. What the tenants were to leave at the end of their twenty-one years' tenancy was stipulated; Benoni Mallett, for instance, on his farm of about 1,600 acres, was to leave 112 acres ready to be sown with wheat, 124 acres of turnips, and 150 acres of land laid down with clover and grass seeds of one year's lying. As for cultivation year by year, tenants were directed not to take 'above 3 crops of corn and one of turnips off any of the lands hereby letten before laying down with Clover & Grass seeds to lye two years'. Those provisions point to a six-course rotation: corn, turnips, corn, corn, grass, grass. But the tenant had the alternative of a four-course: if he kept a ley for one year only, then he must take only two corn crops and one of turnips before laying down to grass again. Indeed he could, presumably, leave out turnips and perhaps substitute a fallow, without violating his covenants. Though not rigid, these were progressive and reasonable conditions of tenure. Such clauses, if heeded by a willing tenantry, or even enforced on a reluctant one, could act as a major stimulus from the landlord towards good farming by his tenants. We have seen that rather similar conditions were imposed at Massingham in 1751; probably Cauldwell introduced this set of covenants into leases during all or most of the years in which he was responsible for drafting them.

The farm in hand continued to flourish under the widowed Lady Leicester. Some figures will give an idea of the scale of operations. In the years 1763-7, to judge by payments for threshing, the home farm produced a yearly average of 779 coombs of barley, 242 coombs of oats, 231 coombs of wheat, plus a mere 18 coombs of rye.[21] To judge by payments for hoeing, an average of about 100 acres was under turnips in each of the same years.

This account of crops serves to refute the old belief expressed thus by Lord Ernle:[22] 'When Coke took his land in hand [i.e. in 1776] not an acre of wheat was to be seen from Holkham to Lynn.

[21] A coomb is four bushels. It is assumed today to equal the following weights: 1 coomb barley or rye = 224 lb.; 1 coomb oats = 168 lb.; 1 coomb wheat = 252 lb.

[22] R. E. Prothero (Lord Ernle), *English Farming Past and Present*, 5th edn. (London, 1936), 217-18.

The thin sandy soil produced but a scanty yield of rye . . . Coke determined to grow wheat.'

In the years 1759-67 the farm made annual average profits of just over £750 before deduction of the nominal rent.[23]

That the farm in hand was still a going concern when Thomas William Coke took over in 1776, is shown by the payments recorded for carrying it on in the audit book for 1775.

The £2,000 a year that Lord Leicester had directed to be paid for finishing the new house at Holkham was duly spent, and the house was finished in 1765.[24] In 1768 the mortgages due to Thurston, Fellows, and Hase were paid off, a total of £5,750. This was done out of the remainder of the share of the personalty of the late duke of Argyll which became due to Lady Mary Campbell on her mother's death, and which she had ceded, in advance, to her father-in-law, Lord Leicester.[25] Lord Leicester had been paid £7,500 in advance and now, on 3 June 1768, £6,505. 4s. 2d. was received through Sir Matthew Lamb.[26] The sinking fund of £3,000 a year for repayment of debt duly came into operation when Lord Leicester died. The general report of Graves, the Master-in-Chancery in Cauldwell's case, certified that Cauldwell had set aside the annual £3,000 from Lady Leicester's death in February 1775 until March 1781.[27]

When Wenman Coke died in 1776, he left unpaid debts of his own of £30,193. 5s. 3d. and left his unsettled estate in Norfolk to be sold to pay the debts that could not be met out of his personalty. His personalty accounted for £15,137. 17s. 10d. and his son, Thomas William Coke, paid the balance and therefore took over the estates Wenman had left to pay his debts. These lands, worth only £6,000 or so in capital value, were apparently lands bought by the earl of Leicester after the date of the codicil to his will and which therefore escaped settlement.[28] Wenman Coke had also been in possession of

[23] Unlabelled account book in Muniment Room of Norfolk receipts and payments 1759-67.

[24] 'Building' account book (Muniment Room).

[25] The duchess of Argyll died on 16 Apr. 1767.

[26] H.F.D., 51, 62. Lady Mary Campbell's longevity rendered her marriage unprofitable to the Coke estate in the long run; from Edward Coke's death in 1753 until her death in 1811 £2,000 a year had to be paid to her. G.E.D., 64(i), ff. 69-70.

[27] G.E.D., 79: Act of 1785, *For Vesting certain Estates . . . devised by the Will of Thomas late Earl of Leicester in Trustees, to be sold; for laying out the Money arising therefrom in the Purchase of other Estates.*

[28] A/B 1782, pp. 74-5; H.F.D., 74a, 75.

estates in Derbyshire and Lancashire worth £6,400 a year net in 1776.[29] At his death in 1634 the main estates of Chief Justice Coke, including the Norfolk estates, descended through his longest-lived son, Henry, while his Derbyshire properties went through his next-born son, Clement, who acquired the Lancashire estate by marriage. Clement's descendants died out in 1727, when Sir Edward Coke of Longford, Derbyshire, died. He left the Derbyshire and Lancashire estates in turn to the two younger brothers of Thomas Coke. Both these younger brothers died, childless, before their eldest brother. Thus in 1754 the estates went to their nephew Wenman.[30] When Wenman came into the Holkham estates at Lady Leicester's death in 1775, he reunited the two groups of estates. But they were not kept together for long. Possibly Wenman, and certainly his son, Thomas William, devoted the income from the Derbyshire estates, and perhaps the Lancashire estates too, to the support of the latter's brother, Edward, though the ownership remained in the hands of Thomas William. The latter eventually maintained what had evidently become a tradition that the Derbyshire estate should go to a younger branch by directing in his will that the estate there should always go to the next remainder-man after the tenant for life of the Holkham estates.[31]

In 1776, then, Thomas William Coke came into a large and, as it seems, well-managed and prosperous estate. He inherited, too, a large house and substantial debts incurred in building that house and in the South Sea fever of 1720. The rest of this book deals with his activities as master of the Grand Estate.

[29] A/B 1782, pp. 74-5.
[30] See the Table of Descent, p. 213, below.
[31] C. W. James, *Chief Justice Coke*, 121-2, 218-21; 1824 L.B., 163-4; H.F.D., 117, copy of will of Thomas William, earl of Leicester.

6

Coke of Norfolk and Agricultural History

THOMAS WILLIAM COKE was an eminent and successful landlord. The Park farm at Holkham was regarded by contemporaries as a model of the best farming practice. The management of his estates was widely praised and admired. Coke's activities as farmer and landlord influenced agriculturists throughout England and beyond. Within Norfolk his prestige was unrivalled.

For many years, however, a still greater place in the history of English agriculture was claimed for Coke. For instance, what is still the most valuable single-volume history of English agriculture, Lord Ernle's classic *English Farming Past and Present*, contains the following passages:

In Thomas Coke of Norfolk the new system of large farms and large capital found their most celebrated champion . . . In 1778 the refusal of two tenants to accept leases at an increased rent threw a quantity of land on his hands. He determined to farm the land himself. From that time till his death in 1842, he stood at the head of the new agricultural movement. On his own estates his energy was richly rewarded. Dr. Rigby, writing in 1816, states that the annual rental of Holkham rose from £2,200 in 1776 to £20,000 in 1816. When Coke took his land in hand, not an acre of wheat was to be seen from Holkham to Lynn. The thin sandy soil produced but a scanty yield of rye. Naturally wanting in richness, it was still further impoverished by a barbarous system of cropping. No manure was purchased; a few Norfolk sheep with backs like rabbits, and, here and there, a few half-starved milch cows were the only livestock; the little muck that was produced was miserably poor. Coke determined to grow wheat. He marled and clayed the land, purchased large quantities of manure, drilled his wheat and turnips, grew sainfoin and clover, trebled his live stock . . . By the use of the new discovery [bonemeal] Coke profited largely. He also introduced into the country the use of artificial foods like oil-cake, which, with roots, enabled Norfolk farms to carry increased stock . . .

In nine years Coke had succeeded in growing good crops of wheat on the land which he farmed himself. He next set himself to improve the livestock. After trial of other breeds, and especially of Shorthorns among cattle and of the New Leicester and Merinos among sheep, he adopted Devons and Southdowns. His efforts were not confined to the home farm. Early and late he worked in his smock-frock, assisting tenants to improve their flocks and herds. Grass lands, till he gave them his attention, were wholly neglected in the district. If meadow or pasture wanted renewal, or arable land was to be laid down in grass, farmers either allowed it to tumble down, or threw indiscriminately on the ground a quantity of seed drawn at haphazard from their own or their neighbour's ricks, containing as much rank weed as nutritious herbage . . .

Impressed with the community of interest among owners, occupiers, and labourers, Coke stimulated the enterprise of his tenants, encouraged them to put more money and more labour into the land, and assisted them to take advantage of every new invention and discovery. Experiments with drill husbandry on 3,000 acres of corn land convinced him of its value in economy of time, in sowing of seed, in securing an equal depth of sowing, and in facilitating the cleaning of the land . . . As with the drill so with other innovations. He tested every novelty himself, and offered to his neighbours only the results of his own successful experience. It was thus that the practice of drilling turnips and wheat, and the value of sainfoin, swedes, mangel-wurzel, and potatoes were forced on the notice of Norfolk farmers. His farm-buildings, dwelling-houses, and cottages were models to other landlords. On these he spared no reasonable expense. They cost him, during his tenure of the property, more than half a million of money. By offering long leases of twenty-one years, he guaranteed to improving farmers a return for their energy and outlay . . . At the same time he guarded against the mischief of a long unrestricted tenancy by covenants regulating the course of high-class cultivation. Though management clauses were then comparatively unknown in English leases, his farms commanded competition among the pick of English farmers. . . .[1]

There is not very much truth in all this. Many of the statements of fact are wrong; practices said to have been introduced by Coke had been long established by the time he came to his estates; the conclusions drawn, even from those statements which are more or less accurate, are usually mistaken.

[1] R. E. Prothero, *English Farming Past and Present*, 1st edn. (London, 1912); 6th edn. (London, 1961), 217–20. The latter edition has excellent introductions by G. E. Fussell and O. R. McGregor. These make several reservations to Ernle's statements about Coke. In particular, Mr. Fussell draws attention to Arthur Young's contrary evidence. It is surprising that Lord Ernle failed to notice that his own remarks about the 2nd Viscount Townshend's agricultural activities in north-west Norfolk were incompatible with his description of that area *c.* 1776.

As we have seen, it is wrong to think of the agriculture of west Norfolk as backward before 1776: it is surprising, in this context, that Arthur Young's earlier writings have been so generally ignored. Young wrote about 1770 that the name of Norfolk was 'famous in the farming world'.[2] He praised much of the agriculture of north-western Norfolk. Again, Young wrote a year or two earlier: 'Half the County of Norfolk, within the memory of man, yielded nothing but sheep-feed; whereas those very tracts of land are now covered with as fine barley and rye as any in the world—and great quantities of wheat besides.'[3] Later, in 1784, Young praised Coke's farming, by pointing out that it was not easy for him to make his farm outstanding 'in the midst of the best husbandry in Norfolk, where the fields of every tenant are cultivated like gardens'.[4]

The greater part of the Coke estates was enclosed before 1776.[5] Substantial areas, especially of common, remained to be enclosed and broken up but much sheep-walk had been brought into intensive cultivation in the first half of the century. Sown grasses and clover were used as a matter of course before 1776 and wheat was grown in substantial quantity at Holkham and elsewhere long before then.[6] Marling had been familiar for a very long time. The elements of 'Norfolk husbandry' were present in north-west Norfolk long before 1776, long leases containing husbandry covenants calling for their use were no novelty, and there were excellent farmers, deploying large capital, holding Coke farms before Thomas William Coke succeeded his father. Benoni Mallett, with his great farm at Dunton, and Carr of Massingham, the management of whose farm gave the first Lord Leicester 'great pleasure to see' in 1747,[7] as it did to Young in 1770,[8] had many equally rich and accomplished colleagues.

It used to be alleged that Coke took up farming because of the accident of the refusal, shortly after he came to the estate, by two tenants at Holkham of new leases at moderately increased rents. The story appeared in Dr. Rigby's pamphlet of 1816 and was

[2] Arthur Young, *Farmers' Tour through the East of England*, ii (London, 1771), 150.
[3] Id., *The Farmers' Letters*, i (3rd edn., London, 1771), 9–10.
[4] Id., *Annals of Agriculture*, ii (1784), 353.
[5] The condition of the estate at that date will be discussed further in a subsequent chapter.
[6] Miss N. Riches very reasonably argued that Young would certainly have discussed any attempt to introduce the culture of wheat in an area previously without it. *Agricultural Revolution in Norfolk* (Chapel Hill, 1937), 94.
[7] Letter to Carr, probably from Cauldwell, 4 Mar. 1747 (?8) in Massingham Deeds, 236.
[8] *Farmers' Tour through the East of England*, ii (1711), 1–6.

elaborated into its twentieth-century form by Mrs. A. M. W. Stirling:

One of the first discoveries made by Coke on succeeding to his property was that the leases in the parish of Holkham, granted by old Lord Leicester, were about to expire . . . In the leases previous to the ones then current, these farms had been let for eighteenpence an acre; in the current leases this had been revised to three-and-sixpence. Coke sent for the tenants, Mr. Brett and Mr. Tann, and offered to renew their leases at the moderate rental of five shillings. Both refused, and Mr. Brett, who, as Lord Spencer remarks, ought to have his name recorded for the good which he unintentionally did his country, jeered at the suggestion, and pointed out that the land was not worth the eighteenpence an acre which had originally been paid for it. This was sufficient for a man of Coke's temperament; he immediately decided to farm the land himself.[9]

The main motive, in fact, was simply that a farm had been kept in hand in Holkham since 1721, a farm which had long been thriving and which was still a going concern when T. W. Coke took over in 1776. Certainly the home farm did grow in size soon after Coke came—which may account for the story of the reluctant tenants.

In 1776 Thomas Tann held the Honclecronkdale farm of 340 acres of arable and marsh on a lease of eighteen years from 1761 at £178. 18s. 0d. (a figure difficult to reconcile with 3s. or 3s. 6d. an acre). He did not take a new lease in 1779, before which date more land had been added to his farm, bringing the rent to £190. 6s. 0d. In 1779-80 the farm was held by William Money at £191. 17s. 0d. The year after, it was in hand, but it was let out again for twenty-one years from Michaelmas 1781 to Jeremiah Sharpe at a rent of £150 per annum. This might suggest that rent per acre had been reduced—and a note in the audit book says the rent had 'decreased' —but it is more likely that some land had been taken from this farm and added to the farm in hand.

In 1776 the more notorious Thomas Brett held the Ostrich Inn with about 40 acres (some of which was arable, some of it valuable marsh) at a rent of £30. 2s. 6d. In 1779 part of a farm formerly held by Henry Winn was given to him and his rent became £220 a year. But by 1787 Brett had disappeared from the list of tenants in Holkham. The inn and some marshes, but no arable, were then held by

---

[9] A. M. W. Stirling, *Coke of Norfolk and his Friends*, 1st edn., 2 vols. (London, 1908), i. 250-1; 2nd edn., 1 vol. (1912), 157.

William Fodder for £23 a year. Evidently the farm that had once been held by Winn, together with the arable Brett had held with the inn, had been added to the farm in hand, the 'Hall farm', or as it soon began to be called, the 'Park farm'. Winn's farm, the 'Home farm'—a misleading name since it was probably never in hand until this time—had consisted of 377 acres of arable, let to Winn, tithe free, at £134. 14s. 0d., a rent, again, far above 3s. or 3s. 6d. an acre.[10] But there is evidence, from much later, to suggest that some of the land added to the Park farm at this time was very poor land and it may well be that a small area, which became a tiny portion of the Park farm, had been valued at 3s. an acre. In 1831 Blaikie, the steward, corresponded about it with R. N. Bacon (by that time these domestic details of Holkham affairs had already begun to loom large in the historiography of Norfolk farming).[11] Bacon was associated with the *Norwich Mercury* and was the author, in 1844, of a prize report to the R.A.S.E. on the agriculture of Norfolk. Blaikie wrote to correct statements made by Bacon in the *Norwich Mercury* and told Bacon that

> The lease of the farm to which you allude expired several years after Mr. Coke came to his property. The land is situated in Holkham, as you state, and you are correct so far—It was occupied by the late Mr. Brett, father of Mr. Willm. Brett of Burnham Overy in the neighbourhood to whom reference may be made. Mr. Brett, senior, rented the farm in question at three shillings per acre, and Mr. Coke at the expiration of Mr. Brett's lease offered him a renewal at the rent of five shillings per acre, Rectorial Tithe included'.

This land could obviously only have been a small proportion of Winn's farm—if it was part of that farm at all—since that farm was let at an average rent of 8s. 6d. an acre when Coke came in 1776; it seems more likely that it was the small amount of arable land Brett had held with the Ostrich Inn.

And although [wrote Blaikie] Poors Rates were merely nominal at that time, Mr. Brett considered the rent set too high. He declined accepting Mr. Coke's offer and resigned the occupation of the farm. Mr. Coke then took the land into his own hands, and has never since let any part of it, either at three pounds an acre, as you state, or at any other rent.[12] . . .

[10] A/B 1776 87.
[11] For Blaikie, see below, chap. 10, sect. i.
[12] Bacon had based this story, as many of the Coke myths have been based, on Rigby's pamphlet of 1816 from which Bacon drew the conclusion that 1776 rents could be multiplied by 10 to give their 1816 value—and then he substituted £3 for 30s.

Mr. Coke planted part of it with Forest trees and has continued the remainder in Arable culture—It is land of inferior quality, and, although it has by means of very superior culture produced abundant crops of Corn in moist seasons, as in the last year, yet if it was to let for a term of years I should not consider it worth ten shillings per acre tithe free, but subject to the operation of the poor's laws. In fact a considerable part of it is not worth five shillings per acres for a term of years. The land in question is situate in front of the Holkham Kitchen Garden and Gardener's house. The road from the hall to the West lodge runs through it ...

Mr. Brett, therefore, was not the ludicrous incompetent he was usually painted, nor was his renunciation of his small holding and its addition to the Park farm an event of any particular significance.

Blaikie reverted to this issue in another letter a day or two later:

As to the land now in the Park, late Brett's farm which we will call the 3/- land. I certainly have in favourable seasons, seen heavy crops of Corn on particular patches of it—but I have also, in dry seasons, seen total failure of Crops upon other parts of that land. There is some Arable land of first-rate quality, which may be worth 30/- the acre, not £3 as you state, immediately adjoining the 3/- land. It is a part of a farm formerly occupied by a man named Tann, and I think it probable this land has in the printed reports been amalgamated with Brett's 3/- land.

Finally, Bacon submitted to Blaikie a paragraph of retraction for insertion in the *Mercury*.[13]

In this correspondence Blaikie firmly denied exaggerated stories current then and since of increases in the Norfolk rents: 'Mr. Coke has greatly increased his Estate in Norfolk, by purchase, since he came to his property, but it is absurd nonsense to say he has increased the value of his Estate from £2,200 to £20,000 by improvements of the soil, since he has been in possession.' This 'absurd nonsense', was based on Edward Rigby's pamphlet of 1816 on *Holkham and its Agriculture*. Rigby, however, though often cited as the authority for this 'absurd nonsense' never made the statement that rents had risen from £2,200 to £20,000 between 1776 and 1816. On one page of his pamphlet he says: 'When Mr. Coke came to his estate at Holkham, the rental was two thousand, two hundred pounds—this was forty-one years ago. The produce of his woods and plantations amounts now to a larger sum.' Three pages further on, he declares that Coke's 'relatively modest rents ... have admitted the total increase of his Norfolk rents to amount to the enormous sum of twenty thousand pounds'. The Holkham

[13] A.L.B., i. 137-8, 144-8.

division of the Norfolk estate is, of course, quite incommensurate with the Norfolk estate as a whole.[14]

The truth is that rents on the estates rose by 1816 to twice their level of 1776. If the rise in prices in those years is borne in mind, the change in rents is not at all startling. The settled lands of the Norfolk estate brought in £12,332 in gross rents in 1776; in 1816 settled land in Norfolk yielded £31,050 a year. In 1785-7 Norfolk lands of the rental value of £2,277 were brought into settlement and more land, worth another £1,513, in 1812.[15] From these figures can be worked out an approximate percentage rise in rent on the Norfolk estates between 1776 and 1816 for comparable areas: it is 105 per cent. Thus rents on Coke's estate in Norfolk were multiplied in the years 1776 to 1816 not by four or ten but by two.

On another occasion Blaikie attempted to correct some of the myths of Coke's rising rents. In about 1831 he wrote to Sir John Sinclair on a proposed appendix to a new edition of the latter's *Code of Agriculture*:

The assertion, that in consequence of Mr. Coke granting leases to his Tenants, His Rent Roll has encreased within the memory of man from £5,000 to £40,000 per Annum is altogether erroneous . . . And it is probable the mistake has arisen from confounding two terms of very different import. These are, 'The Rental of the *Holkham Division* of Mr. Coke's Estate' and 'The Rental of Mr. Coke's Estates in Norfolk'.[16]

Sinclair's figures were probably derived from Kent's remark in his *General View*: 'The Holkham estate . . . has been increased, in the memory of man, from five to upwards of twenty thousand pounds a year . . . and is still increasing like a snowball',[17] but they are further from the truth than Kent's figures, which are themselves inaccurate. In spite of Blaikie's efforts at the time and the later contributions to accuracy of C. W. James and Naomi Riches,[18] the myths about rents and all the other topics were frequently repeated. Miss Riches, indeed, contributed her own mistake: she wrote that Coke's rents increased from £5,000 to £20,000 in the

[14] E. Rigby, *Holkham, its Agriculture &c.*, 2nd edn. (Norwich, 1817), 21, 24. The 1st edn. came out in 1816.
[15] A/B 1776, 1816; H.F.D., 100; Act of Parliament, 52 Geo. III, *c*. 177; H.F.D., 79, 81.
[16] A.L.B., i. 174.
[17] Nathaniel Kent, *General View of the Agriculture of the County of Norfolk* (London, 1796), 123.
[18] C. W. James, *Chief Justice Coke*, 264-5 quotes Young's *Tour* and gives figures for rents in 1741 and 1776. N. Riches, *Agricultural Revolution in Norfolk*, 75, 92-6, noted some of Rigby's errors.

forty years up to 1816—rather than, as Kent claimed, in the span of the memory of a man looking back from 1794 or so.[19]

Where did these errors and exaggerations begin, and how did their truth come to be asserted with such confidence? The person chiefly responsible seems to have been none other than Coke of Norfolk himself. There is no sign that he sought to check false reports of his successes, whatever Blaikie may have done, and there are signs that he contributed to their spread. Very probably Coke relished his great reputation; and in any case great landlords are not necessarily profoundly devoted to literary exactness or historical truth. For many of the years in which he presided at Holkham, Coke was a very old man and when he died in 1842 he was 88. Even in 1816, he was 62 years old. Old men forget facts and remember fancies, especially about the scenes and events of their youth. By the end, Coke had probably come to believe what had been said and written about himself; he may even have done so all along. Possibly it mattered little to him whether what was said was true in detail; he knew himself to be a distinguished agriculturist and an enlightened landlord, and he knew that he deserved applause. But there were more cogent reasons why praise of his achievements should spread and not be discouraged by Coke. Coke was the largest landowner in Norfolk and he normally wielded a dominant political influence in that county. His power inevitably aroused envy and opposition, and his fixed and zealous Foxite Whig opinions inevitably brought more hostility and contestation. Politics were not a major theme in the Holkham sheep-shearings, but even there Coke sometimes proclaimed his views. In 1806, for instance, he declared that he 'always disliked the measures of Mr. Pitt . . . He was compelled to call their attention to the general conduct of that great and able statesman, Mr. Fox, whose wisdom and talents had always predicted the fatal consequences which must arise from the measures pursued.'[20] Coke's political opponents accused him of causing unemployment on the land and high prices for its products. The picture of his fertilizing a desert waste, causing wheat to grow where only rabbits and rye were found before, was evidently called into being in reply.[21] As the over-emphatic counterblasts were

---

[19] N. Riches, op. cit., 128. Miss Riches quotes Kent correctly on p. 75.
[20] *Norwich Mercury*, 28 June 1806.
[21] In the 1950s, it may be remarked, a drawing of two rabbits fighting near Holkham for one blade of grass appeared on a British Rail poster showing Norfolk features of interest.

repeated, Coke, no doubt, came more and more to believe them himself. In any event, many of the stories in Rigby's pamphlet seem to have been repeated by Coke himself to Earl Spencer and others later on.

Edward Rigby was a doctor of medicine and a prominent Norwich 'philosopher' in the sense that word had then. Here is what he says about his pamphlet in his preface:

> The following Paper was read at the Norwich Philosophical Society in December 1816. It was written from notes taken at Holkham, and, obviously at a time when no remark in it could, of possibility, have reference to a contested Election; nor was it ever intended for publication. The late contest for the county has however, brought it forth: the hostility to Mr. Coke, in the course of the election, marked, as it was, with unusual asperity, was chiefly directed against him, as a great landed proprietor, and a distinguished agriculturalist; for the imputed injury done the country by the change he has effected in the system of farming, which was charged with producing various ill consequences, with depriving the poor of employment and rendering corn dear . . .

In the preface to the second edition of November 1817, Rigby wrote again of

> . . . the extraordinary charges, which with unabated hostility, continue to be directed against Mr. Coke and his system, and which, I lament to repeat, are not confined to the ignorant and prejudiced of the lower classes. They are however [he goes on] of easy refutation; a very simple statement will probably satisfy the ingenuous reader, and the most obdurate opposer of Mr. Coke will, I apprehend, be little able to resist positive facts.

It is clear that Rigby was defending Coke from criticism: exaggerated and irresponsible attacks called forth an exaggerated and irresponsible reply.

It is Rigby's essay and the short contribution by Lord Spencer to the *Journal of the R.A.S.E.* for 1842[22] 'on the Improvements which have taken place in West Norfolk' that were chiefly to blame for misleading historians. Nearly every detail of the brightly coloured picture of Coke's work that is reproduced above was originally contributed by one or other of those writers. Whence did their information come? Blaikie denied that he had given Rigby any information at all except on the subjects of the size of farms on the Coke estate and on the cottage occupations there.[23] Rigby speaks

[22] *J.R.A.S.E.* iii (1842), 1–9.   [23] A.L.B., i. 144.

as if his information came from Coke himself. He had 'the advantage of riding with Mr. Coke several hours, two successive mornings, over the Holkham farm in his own occupation, and over another at Warham, occupied by an intelligent tenant; and, as he allowed me to be full of questions, and seemed to have a ready pleasure in answering them, I had ample means of gratification and information'. Coke is quoted as the authority for some fabulous home farm yields: 10 to 12 coombs per acre of wheat; nearly 20 coombs per acre of barley.[24] Coke gave Rigby the story of the reluctant tenants and he is quoted on many other subjects; undoubtedly he was Rigby's main source of information. At the sheep-shearing of 1819 Coke proposed a toast to Rigby, who replied by expatiating again on the poverty of Holkham before Coke came.[25]

At the later sheep-shearings, indeed, there was frequent exaggeration, in compliment to Coke, of the backwardness of north-western Norfolk farming before 1776. At the sheep-shearing of 1820, Mr. Curwen declared after dinner one day, 'I knew Holkham and this part of the county of Norfolk some years prior to Mr. Coke's becoming the possessor of this place . . . The lands . . . were little better than a rabbit warren.' Soon afterwards, Coke himself spoke and repeated the already hackneyed story of the reluctant tenant. Some land which had been taken over from the innkeeper, Coke declared, when he refused to pay 5s. an acre for it, had produced in 1800 or 1801 nearly 80 bushels an acre. 'This shews what a different course of husbandry can effect.'[26] In 1818 Coke had declared that when he began the sheep-shearings 'the land of Holkham was so poor and unproductive that some of it was not worth five shillings an acre'.[27]

Lord Spencer explained in 1842 that in order to inform the R.A.S.E. of what had been achieved in west Norfolk and how it had been done he had 'made the best inquiries I could from Lord Leicester, the only person now living who is able to recollect the former state of this district, and to tell the means by which it has

[24] It is just possible that one field, very heavily manured, produced this yield of barley in one year. But Rigby (pp. 2–3) writes as if yields of this kind could be got from the '3/- land' discussed above, and does not suggest that such yields were extraordinary exceptions. The *highest* yield on *any of the fields* of the Park farm sown with barley in 1951 was less than 13 coombs to the acre and in 1952 less than 12; *the most productive* field of wheat in 1951 yielded not quite 12½ coombs per acre and in 1952 less than 12.

[25] *Norwich Mercury*, 10 July 1819.
[26] Ibid., 8 July 1820.
[27] Ibid., 13 July 1818.

been improved into its present flourishing condition'. He, too, then, relied on information from Coke himself.[28] Moreover, Hillyard, the author of a book, *Practical Farming and Grazing*, published in 1844, quoted Lord Leicester:

> Lord Leicester told me, that on his coming into possession of his estates, in the year 1776, the whole district round Holkham had been let at 1/6 an acre, but was at that time let on lease, of which there were two years unexpired, at 3/- an acre. He offered the tenant a renewal of the lease, at 5/- an acre, which offer not being accepted, he determined to take the land into his own occupation.

Thus the '3/- land' which, as we have seen, can only have been a very small area, had become magnified into 'the whole district round Holkham'.[29]

Thus it appears that Coke himself was largely responsible for the exaggerated accounts of his powers as an impeccable farmer and landlord which have become an almost ineradicable part of writing on agricultural history. He seems to have been reacting to criticisms of the social effects of the style of management of his estates. Such criticisms, if unchecked, could be damaging to his popularity, in itself a valued possession, and to that linked asset, his political influence in Norfolk. The political influence of an English landlord was not a mechanical product of an extensive property; it required, in addition, some measure of goodwill, at least among the articulate section of the rural population. A depopulating landlord was always unpopular and in the early nineteenth century, when population growth was visibly generating menacing social pressures, was arguably dangerous. Coke was anxious to show that the 'new husbandry', even though it rested on large farms occupied by rich tenants, increased rather than diminished opportunities for employment. The most effective argument was that it brought cultivation to land which hitherto lay idle and unproductive. This led to the attribution to Coke of Norfolk of much work that had been carried out before 1776.

In fact Coke was a successful and popular landlord. By encouraging a rich and enterprising tenantry, he contributed significantly, as his predecessors had done, to the development of high farming—

[28] Coke became earl of Leicester in 1837.

[29] C. Hillyard, *Practical Farming and Grazing*, 4th edn. (London, 1844), 348. Hillyard 'often had the honour of being at Holkham with many of our Nobility' (p. 350).

of intensification of arable cultivation in Norfolk. Agricultural production was increased, the prosperity of farmers grew, demand for labour (at least until the beginnings of mechanization) increased. This development, though more slow and less dramatic than the rapid transformation claimed for Coke's influence, was real and important.

7

The estate under Coke, 1776-1816

i. *The structure of ownership and tenancy*

THE estates in Norfolk that Coke inherited in 1776 contained just over 30,000 acres. In 1780 (about the time of the new surveys, on which the acreage figure is based), total rents in Norfolk were £13,118—an average of about 8s. 9d. an acre. The estates still included substantial properties outside Norfolk: Kingsdown in Kent worth about £850 a year, Portbury in Somerset of about £1,200 a year, Minster Lovell in Oxfordshire bringing in about £700 a year in addition to revenue from sales of timber, and Bevis Marks in London worth about £200 a year.[1] The bulk of the Norfolk properties was concentrated in two blocks:[2] one around Holkham in the north stretching southwards for over ten miles and another block of properties further south, separated by about eight miles from the northern group, and including the estates in Massingham, Tittleshall, Castleacre, West Lexham, Weasenham, Wellingham, Kempstone, and Longham. Separated from these two groups were Flitcham on the west and Elmham, Billingford, Bintree, Sparham, and Fulmodestone on the east. Blaikie, Coke's steward, once described the nature of the soil of the Norfolk estates, considered as a whole: 'The soil varies from light dry sand, to strong loam retentive of wet. But the greater part of the land is a friable sandy loam, naturally poor . . . the subsoil . . . is calcareous, and is called clay, marl or chalk, according to the texture.' As Blaikie pointed out, this land was very productive but only under careful cultivation; the high standard of agricultural technique for which north-western Norfolk became famous was imposed by the character of its soil.[3]

[1] A/B 1780, 1757; G.E.D., 77; 'Farms upon the Home Estates Collected' (Estate Office). The London figure is for 1757. There were also the estates in Lancashire and Derbyshire which Coke owned, but from which, probably for Lancashire, and certainly for Derbyshire, he drew no rent.
[2] See Norfolk map, showing the Coke estates.
[3] A.L.B., i. 153.

The nature and value of the soil varied greatly within each estate and even within individual farms. Such frequent changes in type of soil within small areas are characteristic of north-western Norfolk. Holkham Park farm itself illustrates this point very clearly: as Keary reported in 1851, 'The Arable Land is extremely variable, and almost every description of soil met with in West Norfolk may here be seen.' He described 'some good and deep brown loams on a valuable subsoil of brick-earth sand and gravel'—elsewhere 'the land becomes extremely light, resting on a sharp and hungry subsoil' and there were some fields 'rather thin of soil' but on a subsoil of 'marl and clay of a very good description' and so on.[4] On the Wicken farm in Castleacre there was land 'the soil of which is thin and poor, resting chiefly upon chalk' not far from 'a somewhat stiff and retentive loam upon clay'.[5] Some land at Tittleshall was difficult to cultivate because it was too heavy. But in spite of many exceptions it is reasonable to regard the area as one of 'good sand' —good, that is, compared with, for instance, the near-desert sand of the Breckland area around Thetford and Brandon.

Within the areas of Coke ownership there still remained properties belonging to other owners, properties adjoining, and sometimes intermingled with, Coke lands. In some estates all strips and most commons had gone, while in others there remained small pieces; often large areas of common survived. The estates must be examined one by one. But before doing so, one very important point must be made: it is quite clear that there were several areas where ownership of strips had survived but in which actual cultivation of the soil was in no way affected by this survival. The 1779 map shows considerable areas of open-field strips in Kempstone, but a note in the accompanying survey book explains that the implications were purely legal. At that time Mr. Heard farmed 430 acres in Kempstone as a tenant, some in strips; he himself owned another 250 acres and the strips that he owned were intermingled with those he held as a tenant of Coke's. The result was that

> A considerable part of Kempstone was open Fields, but Mr. Heard with the concurrence of Ralph Cauldwell Esqr. Inclosed the same; but had not any regard to the division of Properties, by which means the Baulks are destroy'd, without any Marks being left to the major part of them ... the divisions are therefore marked upon the New plan from the

[4] Keary, i, ff. 2–3.
[5] Ibid. i, f. 223.

The estate under Coke, 1776-1816

Old one which appears (from what remains in the state it was in when it was made) to be very correct.[6]

Coke soon bought all Heard's property,[7] but before then it is clear that the distribution of ownership in Kempstone had no effect at all on field structure or on cultivation. One would expect similar results wherever a farmer's property was intimately mingled with land he held as a tenant—the Kempstone instance was specially noted probably only because there were no marks to show where baulks had been; elsewhere there would be marks to indicate where they had been. The field in Longham called the 'Sixteen Acres' appears in the map of 1779 as divided into nine strips, some of them of less than three roods, others (with ownership consolidated earlier) of three or four acres. Of these strips, Coke owned five, all of which were held by Thomas Hastings as his tenant; the other four were owned by Hastings—the apparent survival of open-field shown on the map would have no real effect. In any case, the legal situation was brought into line with the agrarian when Coke bought Hastings's intermingled freeholds soon afterwards.[8] A similar state of affairs was to be found in Massingham where a field shown on the map of 1779 as divided into strips, some belonging to Coke, some to a tenant, was in fact cultivated by that tenant as a single field: this was the 'Cherry Pightles' of 23 acres, divided, as to formal ownership, into ten parts.[9]

In some townships, however, strips owned by one or more persons who were not tenants were intermingled with strips belonging to the estate—and there one would expect the situation to be more difficult. In Billingford, Thomas Wightman, a tenant, held $10\frac{1}{2}$ acres in the 'Middle Furlong'. His holdings were in eight strips, separated from each other by strips belonging to six other owners. Yet in the field book showing what crops were being grown in each field in tenants' hands from 1789 to the early 1800s, Wightman's $10\frac{1}{2}$ acres are treated as one field under the name 'Middle Furlong' and shown carrying crops in an alternate five- and six-course rotation of the then usual type. Evidently the various owners in the Middle Furlong had reached some sort of agreement; perhaps they had redistributed and consolidated their holdings without changing

[6] 'Plans and Particulars of Norfolk Estates' (Dugmore) (Estate Office).
[7] 'Plans' map and 'Collection of Norfolk farms' (Estate Office).
[8] Ibid.
[9] Ibid., and Field Book 1789-1802 (Game Larder).

the pattern of the ownership of the land, perhaps they had agreed on a common course of crops. However it had been done, the restraints of old-fashioned open field had been overcome.[10] In Weasenham intermingled strips were owned by several different owners and separate people sometimes held strips in the same furlong as tenants of Coke. Even there, as in Billingford, some of those furlongs are included in the cropping book of 1789 and shown there carrying up-to-date rotations; but, significantly, some of them are not. Furthermore, in the enumeration of the ownership and tenancy of lands in Weasenham, many strips held by tenants are listed separately from their other holdings as 'field land', a category to be found, by then, on few, if any, of the other Coke estates. In a lease of 1789, of 571 acres in this parish, to Thomas Sanctuary, 'whole year or Breck lands' are distinguished from the open-field lands he rented, which he was directed to manage in the course 'immemorially used'. Thus in that part of the estates the disabilities of open field survived as well as the formal structure of ownership associated with strips. It is important not to take maps or even 'particulars' of farms too literally; what they appear to show needs corroboration from evidence such as that from leases or cropping books.[11]

In examining the estates in Norfolk the first point to be considered is distribution of ownership of land within the areas in which Coke properties lay. In some places Coke owned unbroken blocks of land; in some, his lands were sprinkled with properties belonging to others—usually small landowners who had not yet disappeared.[12] In 1780, in the Holkham division of the estate, Coke owned 3,796 acres, about 420 of which were marshland, 220 were woods, 100 pasture, and the rest arable.[13] By then, there were no traces of open field, and only 35 acres of common or waste. Not much land was owned by anyone else. In the Burnhams (Burnham Overy, Burnham Thorpe, Burnham Sutton, Burnham Westgate, and Burnham Norton) Coke owned 1,096 acres, nearly all in Burnham Sutton and Burnham Overy: in those two parishes, about 203 acres were held by freeholders and there was a small 8-acre common in Burnham Sutton. Burnham Overy was neatly enclosed, but the map of

[10] 'Plans' map and Field Book 1789-1802.
[11] Ibid. Counterpart of lease in game larder.
[12] There is no doubt at all that small landowners were disappearing in the period of this study from parishes in Norfolk in which the Cokes owned a substantial acreage: no chance was lost of buying them out.   [13] Part of this division was in Wells parish.

Burnham Sutton shows very substantial areas of strips, with other owners' lands intermingled with those belonging to Coke; these strips were evidently soon disposed of, if they had not been already, for in 1793 Arthur Young wrote of Overman's farm there that it was composed of 'large modern inclosures'.[14] In Quarles Coke held on long lease 583 acres, all neatly enclosed. The Wighton division included about 1,800 acres of Coke land, nearly all arable, about 177 acres owned by others and 51 acres of common. At Waterden, further south, Coke owned a solid block of 770 acres, the 'far greater part' of which was arable. Both the Wighton and Waterden holdings were fully enclosed. In South Creake, on the other hand, there were several surviving traces of open field. There Coke owned 675 acres, all of it arable, there were 208 acres of common 'belonging to Mr. Coke' and other freeholders owned 93 acres. In Dunton Coke owned about 1,530 acres, all in one farm, of which 130 acres were pasture or meadow; all this land was enclosed. There were freeholders owning 19 acres, and 75 acres common or waste. In the detached Flitcham estate Coke owned 3,276 acres, of which 438 acres was permanent grassland. There were 66 acres in the hands of independent owners and 78 acres of common land. The whole of Flitcham was enclosed. In Massingham 1,034 acres belonged to Coke, about 50 of which were pasture or meadow and the rest arable. Other owners had 119 acres and there were 135 acres of common. About 100 acres were shown on the map divided into strips—in 1751, as we have seen, that land was apparently subject to open-field regulation of cropping. Castleacre contained about 2,589 acres belonging to Coke, of which about 130 acres were pasture or meadow, the rest being arable. Four hundred and fifty acres or so were owned by others and there was common land of about 160 acres. There survived a field divided into small square sections, rather than strips, the 'West Field'. At West Lexham, Coke's property made up 934 acres, all but 40 acres of which was arable and there were 191 acres of freeholds and 37 acres common. Traces of open field survived: strips in the 100 acres or so of 'Dunham Field'. In Weasenham and Wellingham Coke owned 2,265 acres including 200 acres of grassland; 1,784 acres were in other hands and, perhaps in consequence of scattered ownership, there were substantial remnants of open field, over 200 acres in one block and three or four lesser areas of surviving strips were marked on

[14] *Annals of Agriculture*, xix (1793), 458.

the map and there remained also a very large area of common land amounting to 734 acres. Tittleshall, on the other hand, was wholly enclosed. There Coke property was dominant; he had 3,572 acres and other owners 135 with only 36 acres of common surviving. To the south, at Kempstone, Coke owned 430 acres, about 70 pasture, the rest arable. Other owners had 303 acres of which 251 belonged to Coke's tenant there. There were 105 acres of common. The apparent survival of open-field there has already been discussed. In Longham Coke owned 820 acres: about 300 acres were wood or pasture, the rest arable. Other owners had 251 acres and there were 372 acres of common but only faint traces of field land appear on the map. The small detached estate of Ashill had 267 acres of Coke property, with 336 acres of common adjoining it which was soon enclosed.[15] At North Elmham Coke's 414 acres were all pasture land. Billingford contained 1,359 acres belonging to Coke, 340 to other owners, and there were 781 acres (some in adjoining parishes) over which Coke had rights of common. Of Coke's lands 16 acres were wood and 300 acres pasture. In Bintree 480 acres were Coke's, 80 acres being meadow or pasture, and the other owners had 66 acres. Coke had rights of common over no less than 892 acres. Open-field ownership survived both in Billingford and Bintree. In Sparham Coke owned 1,117 acres, 13 of which were woods and 300 acres grassland, and rights of common over 694 acres; freeholders owned 307 acres. The estate at Fulmodestone completes this survey. There Coke had 1,248 acres, nearly all arable. Other owners had 546 acres and there were 602 acres of common.

Clearly many small landowners had not yet disappeared. This survey shows, too, how vast an area of common land still remained. Except where the legal rights of others had hampered the process, it seems that most sheep-walk land had been brought into full cultivation by the processes described earlier, and apart from the commons, there is little sign of breaking up of sheep-walk after 1776. The remaining commons provided the prizes to be secured by buying out all the holders of common rights or, if that was impossible, under enclosure awards.[16]

[15] By Act of 1785. N. Riches, op. cit., 162.
[16] The account of the estates above is based on survey books and maps in the estate office: 'Collection of Norfolk Farms', 'Plans and Particulars of the Norfolk Estates' (Biedermann), 'Plans and Particulars of the Home Estates', 'Farms upon the Home Estates Collected', 'Plans and Particulars of the Norfolk Estates' (Dugmore) are the titles on the respective covers. The surveys were made in 1779–81. The distribution of ownership is that within

Large farms were prominent in the estates; of about 20,000 acres,[17] more than half was worked in farms of over 400 acres, as shown in Table D.

TABLE D

Size of farms in acres in 1780

5-49	50-99	100-299	300-499	500+
25	5	23	18	18

The largest farm of all was Benoni Mallett's in Dunton. He held 1,530 acres. His arable land covered 1,390 acres, his pasture 31 acres, and his meadows 99 acres. His farm included 2 houses, 8 cottages, and 5 barns. For this farm he paid £680 per annum, tithe-free. His farm was an example of one recently grown larger; from 1769 he had held the Dunton farm of nearly 1,100 acres plus nearly 500 acres previously held by Thomas Rhoades.[18]

Buying up of intermingled lands belonging to other owners took place whenever chances arose. The most striking purchases of this kind were made in Weasenham where no less than 1,272 acres was acquired by the end of 1796. The bulk of this land was bought in 1796 for £25,000: the estate of Richard Jackson of 1,195 acres, worth at least £730 a year.[19] Large numbers of strips in the Weasenham open fields were thus transferred to Coke's ownership, and, after a purchase in 1804 of another 82 acres (for £2,750), there remained only a few acres of field land in other hands: 126 acres belonging to Lord Townshend and 172 to Richard Martin. Thus in Weasenham and Wellingham as a whole the proportion of Coke property to others' property, at first of 2,265 to 1,784, had altered to 3,619 to 430: an excellent preliminary, with over 700 acres of common at stake, to the act for enclosure.[20] Other substantial purchases of this sort were made in Kempstone, where Heard's 243 acres were acquired for £4,500 in 1789; in Castleacre, where 199

the individual units of the survey; the surveyors recorded Coke properties plus intermixed properties and adjacent commons—they did not necessarily use parishes as their units and the figures above do not necessarily state exactly the balance of ownership in individual parishes.

[17] G.E.D., 77. [18] P.R.O., C. 12/572/13, f. 5; G.E.D., 77.
[19] A/B 1797, Private Estate. Weasenham conveyance in possession of Mrs. R. Coke.
[20] 'Collection of Norfolk Farms'; G.E.D., 80; A/B 1804 (account current).

acres were bought before 1782 for an unknown price, 33 acres in 1788 for £700, and 117 acres in 1796 for £3,700. All these purchases involved lands closely intermingled with Coke's: in Castleacre the whole 'West Field' became his property.[21] In West Lexham £4,000 was paid in 1813 for some land, and £1,000 in Sparham in 1814. Savory's estate in Wighton was bought for £2,800 and in 1799 the Peterstone estate (part of the Park farm), which had been leased from the Bishop of Norwich, was bought outright for £4,736. 15s. 0d. and the Snoring foldcourse, similarly leased for many years, was acquired in 1810 for £1,382. 19s. 3d. In all, it seems that not less than £50,000, and probably substantially more, was spent on this sort of internal rounding-off between 1780 and 1816.[22] Many of the purchases were quite small. For example, 1 acre of arable 'intermixed in Abel Ward's farm' was bought for £50 in 1802, and 5 perches in Flitcham for £10 in 1806. Then there was a small landowner at Sparham who exchanged independence for security: in 1807 Richard Wymer sold 'the whole of his Estate in Sparham consisting of a Cottage, Barn and Outhouses with 4a.0r.15p.' for £10 and the promise of 10s. a week for the rest of his life.[23]

How were these purchases paid for? Usually out of income, paid, that is, by the Steward out of the rents he received. But for the larger purchases, other resources were needed. It seems probable that the sale of the Lancashire estate contributed substantially. The Coke properties there were sold between 1790 and 1804, during which years £35,000 or so had to be paid for purchases in Norfolk. Over £110,000 was raised from the Lancashire sales: £5,100 of it went to the Bishop of Norwich, probably to pay for land bought from him, £13,000 went to Gurney and Co., Coke's Norwich bankers, and £36,000 to Down and Co.; very likely some of this money was used to pay for newly bought land. In 1796, £12,000 of the Weasenham purchase money was raised by mortgage.[24]

Apart from this internal rounding-off of the estates, two big purchases brought about an expansion of the areas in which Coke property lay into adjoining areas; these were the purchases of the Warham and Egmere estates.

The acquisition of the Warham estate was a difficult and expen-

[21] G.E.D., 80; A/B 1782-96, (Private Estate and accounts current); 'Plans and Particulars of Norfolk Estates' (Dugmore). See Castleacre Map, p. 44.
[22] Not all the conveyances survive.
[23] G.E.D., 80; A/B 1776-1816, Private Estate and accounts current.
[24] 1823 L.B., 117. Weasenham conveyance.

The estate under Coke, 1776-1816

sive transaction. First a contract was made by Coke for the purchase of the whole Norfolk estate of the late Sir John Turner: the manors of Wells, Warham, Hales, and Warham Ducis; the advowson of the Rectory of Warham St. Mary Magdalen and lands and other properties in the parishes of Warham St. Mary Magdalen, Warham All Saints, Wells, Wighton, Hindringham, Old Walsingham, Stiffkey, and Binham. The purchase price was to be £57,750. All this was recited in the Act of Parliament (of 1785) that was needed to enable this money to be raised. This was 'An Act for Vesting certain Estates in the Counties of Kent and Somerset, and in the City of London, devised by the Will of Thomas late earl of Leicester deceased, in Trustees, to be sold; and for laying out of the Money arising therefrom in the Purchase of other Estates, situated in the County of Norfolk, to be settled to the same Uses.'[25] The whole of the estate which Lord Leicester had owned at the time he made his will, in Norfolk and elsewhere, had been settled by his will and he had made no provision for selling or mortgaging any part of the estate in order to buy land in Norfolk or anywhere else. Hence the need for an Act, which was an expensive procedure, making the legal part of the whole operation cost £1,360. The trustees in the Act, the duke of Portland and Lord Walpole, sold Coke's estate in Somerset to James Gordon Esq. for £31,840, the Kingsdown estate in Kent to Duncan Campbell Esq. for £21,458 and two detached farms in Kent to Mr. Whitaker for £2,729. The total fell short of the amount needed to cover the purchase price for Warham and the legal charges. So the Coke estate in London was disposed of, too, to three different purchasers, for £9,250. After all outgoings, £6,126. 14s. 0d. remained to be laid out, as the Act directed, on the purchase of lands in Norfolk to be brought into settlement together with the Warham estate: the value of land unsettled had, of course, to be balanced by the settlement of an equivalent value. But, rather than buy unwanted lands, Coke preferred to settle some of the estates he held in Norfolk by purchase or by descent from his father, estates which were not hitherto in settlement, since they had been bought after Lord Leicester's will had been made, and accordingly, lands in Holkham, Wells, Wighton, and Castleacre of the total value of £306 per annum were conveyed to the trustees of the settlement.[26] The Warham purchase was a good one, since it was

[25] H.F.D., 79.
[26] Warham Deeds, 144c.

made just before the sharp rise in land values of the war years. The estate appears to have brought in over £2,000 a year soon after it was bought.[27]

The other major purchase was that of the substantial farm at Egmere. This contained 1,190 acres and was let, when it was bought in 1812, at £1,700 a year—it was fairly good land, as the rent suggests. It was bought by Coke from Edmond Wodehouse of Great Ryburgh for £53,400.[28] From the text of the last lease granted by Wodehouse, of 1809, it is clear that the sale of the estate was expected then to take place soon.[29] Once again Coke's purchase involved the delays and expense of an Act of Parliament, for further parts of the settled estate had to be sold in order to pay for it. The act was passed in 1812.[30] Coke's estate at Minster Lovell in Oxfordshire was unsettled and sold; it was valued at £61,019. 12s. 0d. To maintain the value of the settled estate and fulfil the conditions under which part of it had been sold, Coke settled, not the newly bought land, but other lands belonging to him in Norfolk, valued at £61,291. 4s. 6d., which he had bought himself and which had therefore been part of his 'Private Estate' as the estate out of settlement was called. This was by no means the whole of the 'Private Estate' in Norfolk, a fact which gives some measure of the scale of his purchases in his first thirty to forty years at Holkham.

On the face of it, the Egmere purchase was less satisfactory than the one at Warham. The farm was bought with the rent, as set in 1809, at £1,700—making the price paid for it about thirty to thirty-one years' purchase. This rent was soon pushed up, by 1817 it was £2,000, and a lease was then granted fixing this rent until 1838. But the depression of the early 1820s soon made £2,000 an impossible rent, and in 1822 it came down to £1,524, at which figure it remained. Egmere, in fact, had been bought very nearly at the peak of war-time land values—though, of course, a high war-time price was no doubt obtained for Minster Lovell.[31]

The Warham and Egmere purchases and the sales made to pay

[27] A/B 1787-8.
[28] Egmere Deeds, 5.
[29] Ibid., 4.
[30] 52 Geo. III, c. 177, *An Act for effectuating an Exchange between Thomas William Coke Esquire, and the Trustees of his settled Estates.*
[31] High prices for land in the Napoleonic war were due to high prices for farming produce—the high interest rates tended on the contrary to keep the price of land low (though it must be remembered that the security for capital value provided by land made it especially attractive in times of political crisis).

for them were major stages in the process by which the Coke estate became concentrated in Norfolk.

When the valuation was made for the part of Coke's 'private estate' (i.e. of the land purchased by Coke since 1776) that was to be settled in 1812, the valuer, J. Dugmore, wrote to P. A. Hanrott, Coke's London solicitor, to say that he had valued it as any ordinary purchaser would. But the lands in question were far more valuable to the Coke estate: 'I feel no hesitation in saying that from the situation and intermixture of the two Properties, I would, most readily, give 3 or 4 years Purchase for the private Estate more than the amount of my Valuation, rather than it should go into other hands.'[32] That was the point of all Coke's purchases; they were purchases of land of special value to the estate. They were not the outcome of Coke's having large sums of money which he was anxious to invest —far from it. They were, on the contrary, purchases which Coke could not avoid making when the properties came into the market, even if paying for them meant substantial inconvenience. Of course, it was only because of their great value to the estate that these purchases imposed themselves: they were bargains too good to miss. The land market did not deal in a homogeneous commodity; to an individual landowner, land in one district might be far more valuable than an equal area of equally good land somewhere else. The extent of Coke's buying of land was determined by the amount of land (in particular parts of north-western Norfolk) that came up for sale—as a result of all sorts of remote causes—and not by causes directly affecting his own willingness or ability to buy. If he bought no land in any particular year, it was not because he was less willing or able then to buy land than in other years, but because no suitably situated parcels were offered for sale.

ii. *Investment, improvement, and growth in rent*

Substantial sums of money continued to be laid out on farm buildings, indeed, after a relatively economical start, massive sums. From 1776 to 1785 an average of £1,203 was returned each year, out of rents, for repairs and improvements on the farms in Norfolk. This was slightly under 9 per cent of gross rents. Thereafter, the heading 'Improvements' disappears from the audit books. In striking contrast to the state of affairs in the first half of the century,

[32] Letter in H.F.D., 100.

expenses such as marling and claying and draining ceased to attract any contribution from the landlord and came to be paid for entirely by the tenants. Coke of Norfolk certainly laid out more on farm buildings than his great-uncle, Lord Leicester, had done, possibly even a greater proportion of his larger rental income, but he spent much less on improving the tenants' land itself.[33] In 1786–95 an average of £2,654, or nearly 14 per cent of gross rents, was laid out each year. In 1796–1805, £4,135 a year was returned, or about 19 per cent of gross rents; in 1806–15, £4,461 or nearly 13 per cent of gross rents. These figures, which are the totals given in the estate accounts, do not record everything; money was also paid out from the Holkham office and materials were directly provided by it. This money and those materials were estimated as equal in total value to two-thirds of the money allowed to tenants. Many years later, however, the then steward, William Baker, declared the basis of this valuation of the Holkham contribution to be 'fallacious'. The estate office staff itself was evidently not certain how much should be counted as spent.[34] It seems likely that calculations based on this proportion would involve substantial exaggeration of the value of the landlord's contribution to spending on repairs and improvements even though such calculations were made by the estate office and even though they eventually formed the basis of a posthumous boast by Coke, made in his will, that he had spent over half a million pounds on estate improvements. Still, the existence of this supplementary payment of uncertain dimensions at least makes it safe to assert that the audit book totals are substantially less than the true figures.

Other problems are raised by these figures. One which is comparatively unimportant is that certain payments were included in the repair figures, merely because 'repairs etc.' happened to be a convenient heading for them. For instance, the cost of the *un*-improvement of the estate that Coke engaged in after he came to Holkham—growing whins on many of the farms in order to create coverts—appears under 'repairs' in the form of annual payments compensating tenants for their consequent loss of land, until the

---

[33] W. Macro, discussing whether it was worthwhile for a farmer to pay for claying and marling, noted that 'it is now quite out of fashion for the proprietors of lands to bear any part of the great expense that attends claying and marling'. This suggests that the contrast between the conduct of Thomas Coke in the earlier part of the eighteenth century and that of Thomas William Coke later on might be paralleled on other estates. *Annals of Agriculture*, ix (1788), 117–28. [34] 1828 L.B., 85; 1843 L.B., 140.

rent itself was adjusted when a new lease was drawn up. Similarly, a few payments for poor rates, tithes, for collecting cottage rents, providing ale for the cottagers, for court expenses, and for crops left by outgoing tenants were made. In 1791, for instance, payments of £120 entered under 'repairs' were intrusions of this kind. On the other hand, there are recorded, but left out from the total payments for repairs, sums of money for activities such as brick-making, entered under the heading of 'general repairs'. These rather more than balance the excess sums counted in the totals for repairs: in 1791, for instance, £380 was laid out in this way. It seems safe to leave the audit book repair figures intact as they stand in the books.

A much more important problem is the difficulty of distinguishing between spending on the repair of existing structures as distinct from the erection of new buildings. The disappearance of the separate heading for 'improvements' may reflect the cessation of payments to encourage tenants to improve their soil, but it certainly does not reflect an end to the creation of fixed capital. It is probable that about one-half or perhaps more of the money spent or allowed to tenants as 'repairs' was for new building. Wholly new or radically reconstructed farm houses appear as well as large barns, stables, cowsheds, and so forth. Unfortunately the accountant did not feel under any pressure to distinguish the costs of new work from repairs to old. For instance, in 1794 the tenant of the farm at Dunton was allowed £700 for finishing a new farmhouse, converting the old one into four tenements, repairing nine cottages, and building three new ones, and the sum allowed was not broken down. Precision is impossible.[35]

Rents doubled in those forty years. As one would expect, the fastest rise was in the years 1806–16; the settled estate in Norfolk returned rents of £21,404 in 1806 and £31,050 in 1816. Land worth about £1,513 a year was brought into settlement when Minster Lovell was sold. Discounting this accretion, average rents for comparable areas of land rose by about 37 per cent in those years.[36] This is by far the most rapid rise in rents in the years of this study. Individual farmers often had to accept great increases of rent if

[35] A/B 1776–1816.
[36] This calculation involves the assumption that newly purchased land increased in value at the same rate as existing land. All the increase attributable to newly enclosed commons is included in the 37 per cent.

their leases fell in at this time. When a lease of a farm in Castleacre, for twenty-one years from 1793 at £390 annual rent, fell in, a new lease was granted at a rent of £920. The rent of another farm in Castleacre rose from £668 to £1,100 when the lease expired in 1813.

The main reason for the rise in rents during these forty years, and especially in the years of the wars against Napoleon, was the rise in the prices that farmers could get for what they produced.[37] A consequence of this rise in prices, and a contributory cause of the rise in rents in 1806-16 was a further expansion of the area of intensive cultivation. Probably nearly all accessible second-order land had been improved long before the outbreak of the Great Wars against France, with one major exception—those lands where the legal rights of commoners could not be bought out. As prices increased during the wars it became worth while to incur the expenses involved in using private legislation to sweep away the restraints on cultivation imposed by the legal rights of commoners. Hence the numerous enclosure acts that affected the estates between 1806 and 1816, several of which can only have been carried out under the impulsion of Coke and his advisers. They added many acres to the estates. Between 1806 and 1816, £7,260. 18s. 10d. is recorded as paid for Coke's share of the costs of enclosures. Acts were carried through for Weasenham, Wellingham, Sparham, Billingford, Fulmodestone, Bawdeswell, and Foxley (two parishes between Sparham and Billingford), Wells, Warham, Longham, Kempstone, Mileham, and Beeston (two parishes touching on Tittleshall, Kempstone, and Longham), and Stibbard and Ryburgh (parishes adjacent to Fulmodestone). The enclosures towards which Coke paid most were those for Wells and Warham, for which his share was £2,184. 11s. 0d. paid between April 1811 and February 1814, for Weasenham and Wellingham (£1,444. 18s. 5d.), for Sparham and Billingford (£1,294. 8s. 6d.), and for Stibbard, Fulmodestone, and Ryburgh (£1,037. 14s. 11d.).[38] There were other expenses besides these charges, which were for the legal processes of enclosure: once the allotments were made, for example, they had to be fenced. However, the estate acquired substantial areas of new land quite cheaply. It was mostly poor land, of course; its former status as common was not accidental.

[37] See Appendix 5.
[38] A/B 1806-15, accounts current and Stokes's account in A/B 1812 and 1813.

The estate under Coke, 1776-1816

The Longham division of the estates provides a good illustration. There, 116 acres of land newly enclosed from the heath, were added to the Longham Hall farm on its north side. The soil of these additions was characteristically heathy and black, 'very thin and resting upon gravel and a sort of iron-stone sand'.[39] The rent of this heathland was set in February 1814 at 5s. an acre, compared with 20s. an acre for the 'breck land on the north side of the road', which had been enclosed in the eighteenth century, and 25s. an acre for the other land of the farm. Someone, probably John Hastings, the tenant of Longham Hall farm, calculated a yield from the heathland of 3 coombs of wheat to the acre or 4 coombs of barley, compared with 5 of wheat or 6 or barley from the middling lands of the farm, and 7 coombs of wheat and 9 of barley from the best lands of the farm.[40] Probably yields from the heathland could be expected to rise later on. Evidently Hastings soon became alarmed by the expense of reclaiming the heath. Blaikie, the new steward, visited his farm in August 1816. He reported that Hastings was 'a zealous and Industrious tenant, but heart broken by his present undertaking'. Blaikie added a note: 'Mr. Coke has promised to take him by the hand.' What this meant was that Coke agreed to allow him £2 an acre for the heathland he had already pared and burned (as a preliminary to putting it under the plough) or for land which he would subsequently pare and burn. In addition, he was to be allowed a further £2 an acre (just under half the cost) for every acre of land he had already clayed and marled, or would subsequently clay and marl 'at the rate of not less than Forty good cart loads of Clay or Marl per Acre'. Furthermore, Coke agreed to allow Hastings £87. 5s. 9d. towards the cost of enclosing the heath, £89. 7s. 4d. for one-half of the expenses incurred by Hastings in raising new fences on the older part of his farm, and £171. 5s. 6d. for half the expenses of throwing down old fences, drainage, and filling in pits and watercourses, and Coke promised to pay half the expenses of any new fencing or throwing-down of old fences that had to be done in the future. Thus, in this instance, we find the landlord once again contributing substantially to the cost of improving the land. But the allowances were to be paid only on condition that Hastings completed the draining of his farm and brought 'the whole into a perfect state of cultivation'.[41] Coke's financial support was probably due

[39] Keary, i. 258.
[40] Longham Deeds, 716.
[41] 1816 L.B., 98, 135; Blaikie's Reports, 16-17.

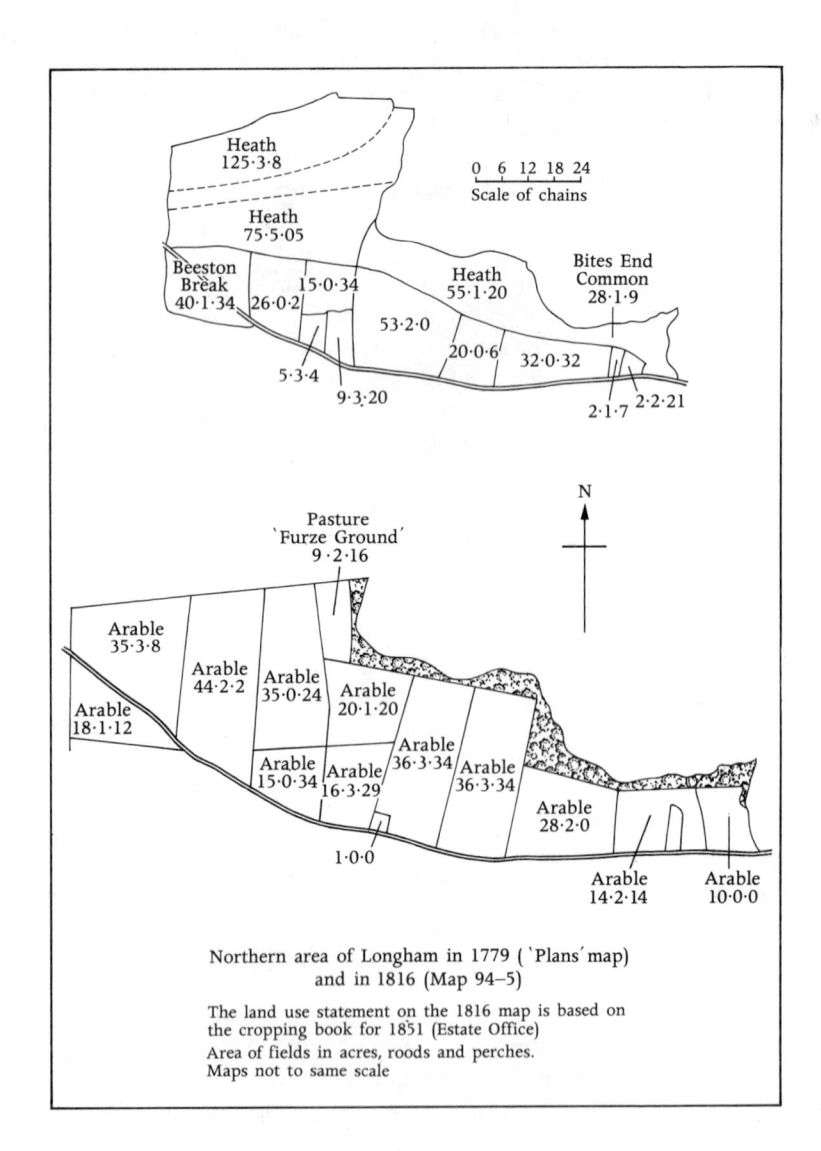

Northern area of Longham in 1779 ('Plans' map) and in 1816 (Map 94–5)

The land use statement on the 1816 map is based on the cropping book for 1851 (Estate Office)
Area of fields in acres, roods and perches.
Maps not to same scale

to the insistence of the tenant on the possibility that cultivating the heath would involve him in loss unless prices for wheat and barley were above 40s. and 20s. a coomb respectively, prices not to be counted on by that time.[42] The subsidies to the tenant of Longham Hall might not have been needed if the reclamation of the heath had taken place a few years earlier, when prices were high.

Such vestiges of open-field arable as remained were disposed of by the enclosure acts but that was much less important than the dividing up of commons. We have seen that only a few remnants of open-field patterns of ownership survived into Coke's time, and that very often they caused no inconvenience at all. The commons were a different matter. Until the enclosure act there survived in Fulmodestone the 'Great Common' of 556 acres; a survey made just after the Fulmodestone enclosure showed that Coke had been awarded 406 acres. Later, there is a report on this land: in one of the Fulmodestone farms there were

... eleven fields which are quite a recent enclosure from the common, are extremely well laid out in large fields with good straight fences, and a convenient driftway running through them from the pasture to the Fakenham road. Claying, draining and high farming have made them productive and valuable, but with anything like second-rate management they would become almost valueless.[43]

In Billingford there had been 327 acres of common; the award map of 1809 shows that it had been partitioned, and maps of 1828 show that Coke's allotment was parcelled out into enclosed fields.[44] The acts for Bawdeswell and Foxley affected 452 acres adjoining Billingford; the Sparham award dealt with 346 acres there; in Weasenham and Wellingham there were 734 acres of common to be partitioned.[45] In all, it would seem safe to guess that Coke must have gained at least 1,500 acres from allotments of common.

Enclosures were of great advantage to owners of substantial property in the parish enclosed. On one occasion, after the war, Coke, by contrast, found himself in the position of a small landowner and his steward protested violently against the result. In

---

[42] See Appendix 3, where calculations about Longham Hall farm, probably from the tenant, are reproduced. They supply interesting information about a tenant's economy. His farming seems strikingly less 'high' than that of the Park farm.

[43] 'Plans and Particulars of the Norfolk Estates'; 'Fulmodestone with Croxton 1811' survey book; Keary, i, 120 (Estate Office).

[44] G.E.D., 77; Map 13/1; B. Leak's survey book, 1828 (Estate Office).

[45] G.E.D., 77.

1830, at the time of the Kettlestone enclosure (a parish adjoining Fulmodestone), Coke claimed allotment of common in return for common rights attached to 70 acres he owned there. The surveyor alloted him less than 3 acres of common.

What makes the case worse [the steward complained] is Mr. Coke's allotment is . . . along two sides of an ancient Enclosure of Mr. Coke's in the occupation of Mr. Applegate—Mr. Coke and his Tenant have recently received notice to raise a fence along that line, which, judging from the appearance on the Map, will exceed 400 Yards in length. The expense of removing the present fence and raising a new one in the line prescribed by the Surveyor, will amount to more than twice the value of the half-acre of land.

Evidently one allotment included a narrow strip alongside existing Coke land.[46] That enclosure was in a parish where Coke property was small; where it was preponderant all enclosures had been carried through in the years 1806-16.[47]

iii. *Leases and husbandry covenants*

William Marshall asserted that 21-year leases were falling out of favour in Norfolk at the beginning of the 1780s.[48] The Coke estates did not share in any such reaction and 21-year leases continued to be the rule on them. In 1788 there were thirty-nine 21-year leases in operation and 16 for 18, 19, or 20 years, 5 for 16 or 17, 9 for 11 to 14 years, 3 for 11 years, and 2 for 7; the last categories seem to provide some evidence to support Marshall, but, on the other hand, terms shorter than 21 years were probably often the result of delay in drafting and drawing up the actual lease document after a new rent had been fixed and an agreement made for a 21-year period. Furthermore, when Coke confirmed the disputed lease agreements made by Cauldwell, the leases were recorded as being for the years remaining of the term originally agreed between Cauldwell and the tenant: Benoni Mallett's agreement was for 21 years from 1769, but in Coke's time his tenure was recorded as 11 years from 1779. In 1816 there were eighteen 21-year leases, 8 for 18, 19, or 20 years, 3 for 16 or 17, 5 for 12 or 14, but none for shorter

[46] 1830 L.B., 38.
[47] Earlier enclosures significantly affecting the estate (Ashill, 1785, and Bintry and Twyford, 1795) were of parishes where Coke property did not predominate. Riches, op. cit., 162. A/B 1796 (accounts current).
[48] William Marshall, *Rural Economy of Norfolk* (London, 1787), i. 67.

terms.[49] Many farms are recorded as being held 'for — years from Michs. 1813', for example; perhaps there was some wavering in the belief in the advantage of 21-year leases at that difficult time. After 1816, Blaikie, the steward, normally employed 21-year leases. Arthur Young wrote that Marshall was correct, but he exonerated Coke from any back-sliding over length of leases.[50]

Until about 1800, the husbandry covenants in the leases continued to be based on the assumption that a six-course rotation including three corn crops in six years and a two-year ley, was the most desirable. This doctrine was laid down with increasing clarity and firmness. (Cauldwell's statement of it, as we have seen, left room for many variations.)

There are two leases surviving from March 1782. Their provisions are not identical: one was based on an agreement made with the tenant by William Wyatt, on behalf of Coke; the agreement for the other was signed for Coke by Edmund Waller. This is the first sign of what appears to be the fact that outside valuers were being employed to let farms, and that this responsibility had been taken away from the steward at Holkham. It helps to explain the inconsistencies that occur in the leases of the years between Coke's succession and 1816, what Blaikie in 1816 called 'incongruities in the former leases'.[51] In 1791 £241. 12s. 0d. was paid to Messrs. Kent, Claridge & Co. for valuing different farms,[52] and on 12 February 1791, Nathaniel Kent signed an agreement with Thomas Hastings for the Longham Hall farm.[53] The agreement was on a printed form on which Coke's name and Kent's name were filled in in ink as well as the date and the tenant's name; this suggests that this was a form provided by Kent rather than by Coke—indeed, that it was possibly a form used by other valuers than Kent. Messrs. Kent, Claridge & Co. were paid another £75. 10s. 11d. in 1795 for 'valuing different farms'.[54] Certainly in the 1790s, then, and probably in the 1780s, too, outside agents were being used for letting farms.

The leases of March 1782 were for a farm in Castleacre (355

[49] A/B 1816.
[50] Arthur Young, *General View of the Agriculture of the County of Norfolk* (London, 1804), 48.
[51] 1816 L.B., 134.
[52] A/B 1791 (accounts current).
[53] Longham Deeds, 715.
[54] A/B 1795 (accounts current).

acres for twenty years from Michaelmas 1781 at £200 a year) and for the Abbey farm in Flitcham (914 acres for thirteen years from Michaelmas 1781 at £450 a year). The Castleacre lease forbade the tenant to sow more than three crops of corn in succession and after the first or second of such crops, he was directed either to 'summer-till such land one Year or take a Crop of Turnips therefrom', and the third crop of corn was to be undersown with grass seeds for a two-year ley. If the land were laid down for only one year then only two corn crops were to be taken before the next ley. The Flitcham lease contained similar provisions except that the tenant was directed to take a turnip crop after his second corn crop; he was, in theory, denied the alternative of a fallow and the choice of whether to break the sequence of corn crops after the first or second year. Leases of 1782 in Tittleshall, 1783 in Wighton, and 1785 in Flitcham had the same provisions as the Flitcham lease of 1782. Sanctuary's lease of 1789 of his farm in Weasenham, on the other hand, gave the option of a fallow instead of turnips and also cleared up the verbal anomaly of speaking of three 'successive' crops of corn or grain, the second of which was to be followed by a turnip crop. In this lease, there appeared an innovation: it was directed that at least 12 lb. of clover seeds should be used to the acre when land was laid down. In the 1790s new provisions appear, evidently drafted by Kent. In February 1791 the agreement mentioned above was made for the Longham Hall farm. The provisions exactly followed those printed by Kent in his book as the 'substance of the contracts which subsist between Thomas William Coke Esq., and his Tenants'.[55] The tenant was to 'endeavour as much as possible to adhere and conform to the course of cropping all his arable lands under six shifts or equal portions'. The course laid down was: (1) turnips or vetches fed off with sheep, (2) spring corn, (3) and (4) ley, (5) wheat, and (6) spring corn. The tenant, it is worth noting, committed himself to agreeing 'to any exchange of land that may be proposed, having other land of equal quantity or value laid to him in lieu of what he may be required to give up'. If the sown grass failed in the second year, the tenant could break it up after only one year, but he should then take only one crop of corn before the next crop of turnips (making a four-course rotation) and he was to bring it 'round again as soon as possible under the regular course of six shifts before stipulated'. This agreement did not lay down

[55] N. Kent, *General View of the Agriculture of the County of Norfolk* (1796), 223–5.

The estate under Coke, 1776-1816

the quantity of grass seeds to be sown. Similar covenants are to be found in a 9-year lease of 1797[56] but in this lease, at least 12 lb. of clover seeds and 1 peck of rye grass were stipulated for the leys.

All Coke's leases thus turned on a six-course of the type set out in Kent's provisions above, with variations. A major innovation appeared in 1801. In October a lease was made of the Waterden farm (815 acres for twenty-one years at £600 yearly rent). In it the tenant was forbidden to take 'any more than two successive crops of any sort of corn, Grain or Pulse from any part of the demised Arable lands and one of such two successive Crops shall always be of Pease or Vetches so that two White Straw Crops (that is) of Barley Oats Wheat or Rye may not in any case ... follow immediately one after the other'. 'No two white straw crops one after the other' became a central Coke slogan. It was repeated again and again in the remaining years of Coke's long life and Coke himself fervently proclaimed the doctrine as a basic and essential principle of sound husbandry. In a period of rapidly rising prices for grain, it is striking to find tenants not merely forbidden to grow a larger acreage of corn than before, but even directed to reduce it—to grow two straw crops in six years instead of three. For the new clause was fitted in to the six-course rotation, with a two-year ley and two green crops, not into a four- or even into a five-course. A four-course would allow one-half straw crops; a five-course with this prohibition, two-fifths; the new six-course allowed only one-third. The drafting of this lease was careless: the tenant was directed to sow winter corn on the broken-up ley and in the year after he was to sow 'that land' with 'Peas or Vetches or with Summer Corn' —which obviously contradicted the two white-straw crop prohibition.[57] This lease ordered that at least 15 lb. of clover seeds and 1 peck of rye grass seeds were to be used for each acre when land was laid down.

A lease of 1804 (Godwick farm, Tittleshall, 566 acres for 20 years at £366 per annum) ignored the principle of the Waterden lease of 1801 and prescribed a six-course: (1) turnips or vetches, (2) spring corn, (3) ley, (4) ley, (5) winter corn, (6) spring corn, which allowed one-half of the land to be under cereals and permitted successive

[56] 11 Oct. 1797, William Mason, Sparham.
[57] Blaikie later referred to leases of this type as 'the ill explained six courses' and 'the unintelligibles'. Blaikie's Reports, 1, 7.

cereal crops.[58] But leases of 1808 and 1810 revert to the prohibition of succeeding straw crops.[59]

A new event is a 21-year lease of October 1815 to Mitchell Forby of Tittleshall. This calls for the famous Norfolk four-course rotation: (1) turnips, (2) barley, (3) grass, (4) corn.[60] Possibly this was an early lease of Blaikie's time backdated to 1815. Certainly it is the first lease to call for a four-course rotation, a fact of some interest. The four-course rotation is still regarded as an innovation of Turnip Townshend's, or at least it is believed that he extended its use. Miss Riches says 'it is undoubtedly true that Townshend popularised the idea' and Mr. Sanders declared 'it was Townshend, however, who ... evolved, on his own estate, the Norfolk four-course rotation, on which a very high proportion of the lighter lands of this country was farmed for nearly two hundred years.'[61] One would expect any such 'popularisation' to involve the spread of the four-course on the Coke estates, which lay north and south of the Townshend properties, and one would assume that the 'high proportion of the lighter lands' included a high proportion of Coke farmland. Yet there is little evidence to show that the four-course was firmly established in the latter half of the eighteenth century and much evidence to suggest that it was not. No doubt it could be found quite often on individual fields—as we have seen, the leases of Cauldwell's time and those of the 1780s permitted it—but it was not normal practice, a virtually inevitable rotation, as it subsequently became. It seems rather to have become firmly established in the later years of the Napoleonic war. After his tour through the east of England, Young declared that the course chiefly adopted by the Norfolk farmers was (1) turnips, (2) barley, (3) clover, or clover and rye grass, (4) wheat[62] but he then says that the grass may extend from one to four years. However he writes, apparently of Carr's farm in

---

[58] A penalty of £50 per acre is stipulated in this lease for land cropped out of order, a provision that could almost certainly never be enforced.

[59] Flitcham leases to George England and H. and W. Burrell. Arthur Young printed in his *Farmers' Calendar*, 6th edn. (London, 1805), 610-1, what he called 'The new covenants in letting the farms of T. W. Coke Esq. MP.' I have not seen at Holkham any leases containing these provisions, but very likely there were some. The covenants call for one-third of the arable lands of a farm to be under sown grass and include a prohibition—without written permission—of successive crops of corn, grain, pulse, rape, or turnips for seed.

[60] Game Larder.

[61] *Agricultural Revolution in Norfolk*, 82; H. G. Sanders, *An Outline of British Crop Husbandry* (Cambridge, 1939), 23.

[62] A. Young, *A Farmer's Tour through the East of England* (London, 1771), ii. 156-7.

Massingham, that the course 'here' was (1) turnips, (2) barley or oats, (3) clover one year, (4) wheat.[63] This is the four-course. Yet in the years 1789-1802, the four-course was followed on none of the fields of this farm—by that time, it is true, it had a different tenant.[64] Nathaniel Kent suggested that the four-course was followed only in 'the best parts' of Norfolk and he implies that a five- or six-course with a one- or two-year ley and three corn crops in each five or six years was usual in the area of the Coke estates: i.e. rotations in the form of (1) wheat, (2) barley, (3) turnips, (4) barley, (5) and, perhaps (6) ley.[65] Young repeated, however, much of the substance of his earlier assertion when he wrote in his *General View*, published in 1804: 'In West Norfolk, the predominant principle which governed their husbandry at that period [at the accession of George III] as well as ever since, was the carefully avoiding two white corn crops in succession. Turnips were made the preparation for barley; and grasses that for wheat or other grain.' The conditions of leases since 1760, as we have seen, imply that Young is wrong in his statement about farmers avoiding successive straw crops; only after 1800 did Coke leases begin to call for such a system of cropping. The rotations Young quoted for Coke farms in this volume are none of them four-course rotations, and his own details of courses of crops suggest that what he once calls 'the old four-shift' was an exception rather than the rule when he wrote.[66] It is possible that the four-course prevailed in the 1760s and 1770s, disappeared for a few decades, and reappeared later, but it seems very unlikely—even though leases of the 1760s, 1770s, and 1780s permitted the four-course, while those of the 1790s specifically called for the six-course. At least one can say that the four-course was not accepted as desirable by the landlord in the 1790s. More evidence will be brought to bear on the question of the chronology of the four-course in a later chapter, as well as in the next section of this chapter.

iv. *Crop rotations employed by tenants*

The major piece of evidence bearing on what tenants did with the land they farmed is the 'Field Book of the Estate of T. W. Coke Esq. in the County of Norfolk'.[67] It shows, for the years 1789 to

[63] Ibid., 3.   [64] Field book (Game Larder).   [65] *General View* (1796), 32-4.
[66] A. Young, *General View of the Agriculture of the County of Norfolk* (London, 1804), 193-218.   [67] Game Larder.

about 1802, what crops were grown each year on the various fields of the farms of Coke's tenants in Norfolk. A 'memorandum' at the beginning of the book claims that 'The benefit which results from a Check of this kind, upon the Tenants of a large Estate, is very great, as it fully guards against any improper course of Cropping, and by shewing what has been the past system of husbandry upon any particular Farm, a Landlord or Agent is the better enabled to judge what regulation to lay it under in future.' It is difficult to know how far this field book should be trusted. Most of the information it records must have come from the tenants themselves. Even if one can assume that the information they gave was accurately recorded in the estate office, there remains the question of how far tenants would give accurate reports on what they were doing. It is uncertain what checks were made on tenants' veracity; without checks of some kind, the tenant might easily be tempted to claim that he was doing what his lease said he ought to be doing, irrespective of whether he did or did not pay great heed to his covenants. But the fact that the book shows quite frequent breaches of the covenants that the tenants had agreed to speaks in favour of the trustworthiness of the record. In any case, we may assume that the book shows either what tenants were doing with their land or alternatively what they thought their landlord would expect them to be doing. Probably it does show what farmers were doing and if this is so, it is very valuable.

This field book confirms plainly that the four-course was not accepted practice. And Young's statement, repeated a second time in his book of 1804, that west Norfolk farmers have with 'singular steadiness . . . adhered to the well-grounded antipathy to taking two crops of white corn in succession' can be flatly contradicted.[68] Blyth of Massingham followed the principle more effectively than most others, yet of his 29 fields, 8 were sown at one time or another with 2 straw crops in succession—barley or oats after wheat—in the years 1789-1802. Arthur Young, in fact, was confusing what was happening with what he felt ought to be happening. Thomas Sanctuary had 25 arable fields on his farm at Weasenham. Only one of them escaped 2 successive straw crops in 1789-1805, and 16 of

---

[68] A. Young, *General View*, 364. 'This', he says, 'is talked of elsewhere, but nowhere so steadily adhered to as in this district.' Even if Coke's farmers were eccentric exceptions to a west Norfolk rule—which is hardly probable—they would together form a very large exception.

them were so cropped more than once. Out of 14 arable fields on the exceedingly light land of the Harpley Dam farm at Flitcham, 9 took 2 straw crops one after the other in the period from 1789-1802. At Waterden, William Hill sowed one straw crop after another on his 23 arable fields on 28 occasions in the years 1789-1801. As for the four-course, its other salient feature—apart from avoiding successive corn crops—is the one-year ley. This appears in the field book much less often than leys for 2 years or more. Of the farms just referred to, the Massingham farm nearly always carried sown leys for 2, or sometimes 3 years; if a ley lasted only for 1 year, the next year was usually devoted to a bare fallow and there were only 5 one-year leys which were not followed by a fallow in 29 fields over 13 years. Sanctuary of Weasenham had only 8 one-year leys in 25 fields in 17 years; the Harpley Dam farm had 8 in 18 fields in 13 years. Hill at Waterden had only 3 one-year leys in 23 fields in 17 years.[69]

The great Harpley Dam farm at Flitcham may be used as an example of the balance of crops on a light land farm. It contained 1,015 acres of which 61 acres were permanent grass and 953 acres arable (70 years before, it was all sheep-walk). A schedule of the crops on the arable in the years 1789-91 and 1800-2 is given in Table E.

TABLE E
(figures to the nearest acre)

	1789	1790	1791	1800	1801	1802
Wheat	101	89	58	143	171	64
Barley	193	306	364	184	223	173
Oats	36	16	..	..	18	43
TOTAL CEREALS	330	411	422	327	412	280
Turnips	205	161	132	194	131	166
Ley	418	279	331	398	363	385
Peas	..	44	..	34	18	..
Fallow	..	58	68	..	..	43
Vetches	..	..	..	..	29	36
Unknown (?Coleseed)	..	..	..	..	..	43
TOTAL ARABLE	953	953	953	953	953	953

[69] I have not counted one-year leys followed by fallows.

As the table shows, the cereal crops at the Harpley Dam farm in these years never covered as much as half the arable area, as they would have done under a four-course. The proportions of crops in these years would fit a five-course rotation: (1) barley, (2) ley, (3) ley or fallow, (4) wheat or oats or barley, (5) turnips. Four, five, six, or seven straw crops were taken from each field in the fourteen years 1789-1802. The proportions of the whole farm devoted to various purposes remained reasonably constant during these earlier war years and, though the amount of wheat grown slightly increased, there was no sign of increasing production of grain. The succession of crops in the individual fields was erratic—no one rotation was rigidly followed. The wheat grown was nearly always taken on broken-up ley or after a fallow, but there are instances of wheat coming after turnips or vetches. The barley crops were preceded by a variety of crops: wheat, oats, turnips, peas, or by a ley, and on one occasion two barley crops were taken in succession. A four-course of (1) oats or wheat, (2) turnips, (3) barley, (4) ley was employed on only one field in the years 1789 to 1800.

The Longham Hall farm was another on poor light soil. Hastings, the tenant there, had between 350 and 400 acres of arable and 108 acres of permanent grass, of which 30 acres was broken up in 1793 and 1794. This was a much higher proportion of grass to arable than that of the Harpley Dam farm, and helps to explain the more severe cropping at Longham, shown in Table F.

TABLE F

(figures to nearest acre)

	1789	1790	1791	1800	1801	1802
Wheat	74	66	53	96	56	116
Barley	97	78	118	50	113	53
Oats	10	26	..	27	20	20
TOTAL CEREALS	181	170	171	173	189	189
Peas	..	..	..	7	..	..
Vetches	..	..	..	..	10	..
Turnips	68	66	68	63	43	106
Ley	109	70	89	145	146	93
Fallow	..	52	30	..	..	..
TOTAL ARABLE	358	358	358	388	388	388

On this farm in Longham, too, the straw crops did not cover one-half of the arable acreage in any year; they were more than balanced by fallows and fodder crops. There is, again, no sign of an increase in acreage devoted to producing corn, though the proportion devoted to wheat grew. The wheat grown was usually taken after ley or fallow—although it was once sown after oats and once after barley. The barley crop almost invariably came after turnips or wheat, but barley on one occasion came after oats and once on ploughed-up ley. This comparative regularity in the management of barley and wheat together with the steady over-all balance of crops would lead one to expect a regular system of cropping applied to each field. Indeed, the proportion of crops grown, with more evenness between leys and turnips, would fit the four course. In fact, however, the cropping of individual fields was most erratic. In some fields as many as nine straw crops were taken in fourteen years. In one field there were grown in four years, following a one-year ley, crops of (1) wheat, (2) oats, (3) wheat, (4) barley, and after turnips the next year, another crop of barley was taken. Another field went nine years without a ley or fallow, growing six corn crops and three of turnips. This state of affairs shows how valuable insistence on the simple slogan of no two successive white straw crops might be: the farmer, already balancing his total crops with care, would be compelled to make his balance go with a sound rotation for each field.

Next the state of affairs on the celebrated Massingham farm is shown in Table G. Some of the farm was poor light soil, some was

TABLE G

(figures to nearest acre)

	1789	1790	1791	1800	1801	1802
Wheat	109	100	110	121	145	50
Barley	250	244	225	174	154	174
Oats	24	11	22	..	..	23
TOTAL CEREALS	383	355	357	295	299	247
Turnips	218	199	169	154	173	207
Ley	226	192	169	252	207	239
Peas	..	32	49	..	..	34
Vetches	..	..	..	26	23	..
Fallow	..	49	83	33	25	33
Unknown	..	..	..	..	33	..
TOTAL ARABLE	827	827	827	760	760	760

good brown loam. The arable land consisted of 827 acres, of which 5 were laid down to permanent grass in 1798 and 1799; 61 acres more escaped from the cropping book when they became the tenant's own property—exchanged for some of his own land. (The land the tenant gave in this exchange was intermingled with Coke's and had been recorded in the cropping book as if the whole of the relevant fields had been held from Coke.)

The cropping at Massingham appears very similar to that of the Harpley Dam farm at Flitcham. On this farm, as on the others, fallows were still by no means unknown—they were frequently used as preparation for wheat. The Massingham tenant, for instance, fallowed the 'New Marled Break' for one year, after a two-year ley, before taking a crop of wheat from it. Successive straw crops were rare on the Massingham farm: there are only two instances after 1795 and none after 1798. On the worst lands of this farm—the large 'breaks' adjoining the common—the cropping was notably cautious and successive straw crops were never taken. The sequence of crops on those fields is shown in Table H.

TABLE H

	Hempit Break 49 acres	Limekiln Break 34 acres	Batten Break 50 acres	Boat Corner 50 acres
1789	turnips	barley	ley	turnips
1790	barley	ley	ley	barley
1791	peas	fallow	turnips	ley
1792	turnips	wheat	barley	fallow
1793	barley	turnips	fallow	wheat
1794	ley	barley	wheat	turnips
1795	ley	ley	turnips	barley
1796	wheat and oats	ley	barley	ley
1797	turnips	peas	ley	ley
1798	barley	turnips	fallow	wheat
1799	ley	barley	wheat	turnips
1800	ley	ley	turnips	barley
1801	wheat	ley	barley	ley
1802	turnips	peas	ley	ley

The sequence on the four worst fields at Massingham is far removed from the four-course Young ascribed to this farm in 1770. In fourteen years only four or five straw crops were taken from each field: the four-course would have produced seven, and even a five-course, with a two-year ley, would probably have produced six on one of the fields.

Individual farmers cropped their land in widely differing ways. The basic elements in their rotations were, however, reasonably constant: corn crops alternating with turnips, sown grasses, generally for two-year leys, and bare fallows. The corn crops were mainly barley and wheat, with barley easily the more important, together with some oats but virtually no rye. (The almost complete disappearance of rye since the early eighteenth century is notable.) No steady rhythm of alternation can be detected among the various farms of the estate, or even among the fields of individual farmers, but the predominant basic types of rotation, with many variations, were these: (1) wheat, (2) barley, (3) turnips, (4) barley, (5) ley, (6) ley; (1) wheat, (2) turnips, (3) barley, (4) ley, (5) ley; (1) wheat, (2) turnips, (3) barley, (4) ley, (5) fallow, but only rarely can one find any single rotation being consistently pursued.

There are other examples of quite fierce cropping, by late eighteenth-century standards, besides those already quoted from Hastings's farm in Longham. At Godwick farm, in Tittleshall, the tenant, J. B. Branford, took a crop of wheat, after a bare fallow, and then followed the wheat crop with three successive crops of barley. At Wighton, Robert Beeston took four corn crops in five years from one field, and three successive corn crops from other fields on two separate occasions. He was thus breaking the clear provisions of his lease of 1783, under which he should have followed either a six-course rotation, with no more than two successive straw crops, or a four-course without any successive straw crops.

This raises the question of how far lease provisions were being observed. Certainly, breaches are to be found quite often, a few of them serious. Yet the general conduct of the farmers was largely in conformity with the tenor of their leases and one would expect the more strict provisions of the nineteenth century to have some effect on their conduct.

The planting of two successive straw crops—generally barley after wheat—was common, even normal, practice on the estates at the end of the eighteenth century. The new provision in leases of the early years of the nineteenth century categorically forbidding successive straw crops is in striking contrast to the habits of Coke's tenants a year or two earlier. Of course, tenants may not have felt themselves constrained by the new clause to act against their own inclination; it may be that the notion of the grave danger involved, especially on light land, in successive crops of wheat, oats, barley, or rye, abruptly became generally accepted in about 1800 so that the farmers would have acted on it soon anyway, irrespective of their leases. Against this, Robert Beeston sowed wheat in 1804 in four fields at Wighton that had just yielded a crop of barley. Thomas Sanctuary of Weasenham sowed barley after wheat on one field in 1802 and again in 1804. William Mason of Sparham sowed barley after wheat in 1803. The tenant of the Harpley Dam farm did likewise in 1802. All these farmers, therefore, would have been affected by the new clause in leases.[70] The conduct of William Money Hill, the tenant of a farm at Waterden, shows how the new type of lease might affect farmers. He had been a confirmed believer in sowing two successive straw crops, in particular in sowing either oats or barley after wheat. But he took a 21-year lease in October 1801 which forbade successive straw crops. His crops are recorded in the field book as late as the year 1805 and he did not again sow one straw crop after another. The last occasion on which he did so was when he sowed barley after wheat in two fields in the spring of 1801. It is tempting to conclude that the new provision decisively influenced tenants' habits—though it would not be absolutely safe to do so. At least, though, it is certain that the new doctrine sharply contrasted with the existing habits of the tenants in the 1790s.

Coke himself thought that his leases were the main reason for the improvements on his estates. On 10 April 1812 he wrote to Arthur Young in his characteristic style of emphatic self-congratulation,

> I am proud in thinking that I have the best tenantry in the Island; that I have the best cultivated estate and not a single farm unlet. What is all this owing to but the confidence which exists between my tenantry and myself, and holding out encouragement to them, by granting leases of 21 years, whereby they are enabled to lay out their capital to advantage, and improving my estate under good and liberal covenants; which secures my property from being injured, and at the close of the lease leaves it in better condition than when we entered into the agreement?
>
> Under such an understanding I have lived to see more than double, I believe I might say with truth nearer treble the quantity of corn grown upon the same given number of acres, by tenants opulent and happy, and *willing two years before the expiration of my leases* to come to an agree-

[70] Some tenants' crops are recorded for a year or two longer than others in the field book.

The estate under Coke, 1776-1816

ment to give me a great increase of rent, which keeps it rolling like a snowball.

We have ample means within ourselves to become in a very short time an exporting country, instead of labouring under the disgrace of trusting to foreign agriculture; which we are encouraging by our own folly, and withholding from our own people. But in order to accomplish this most desirable of all objects, gentlemen must give themselves the trouble (with me I will call it pleasure) of attending a little more to their own concerns, and be more liberal with their tenantry in granting leases, before this object can be attained.[71]

To conclude this survey of the estates in Coke's first forty years at Holkham, Table I gives a statement of the total crops grown on the estate in the period 1790-7.

TABLE I

	1790	1791	1792	1793	1794	1795	1796	1797
Acres of wheat	3,246	3,487	3,575	3,533	3,157	3,537	3,290	3,208
Acres of lent grain (oats, barley, peas)	8,593	7,620	7,874	7,384	7,738	7,789	7,444	7,116
Acres of turnips	3,900	3,843	4,749	4,557	4,285	4,403	4,838	4,117

The estate contained about 30,000 acres so that these figures confirm the suggestion above that very cautious proportions of crops were being grown by the tenants on their farms.[72] There is little evidence to show how far Coke himself intervened in the management of the tenanted portions of his estates. What is certain, however, is his direct interest in the farm at Holkham, and in the sheep-shearing meetings held there, and these are examined in the following chapter.

[71] B.M. Add. MS. 35131, ff. 338-9.
[72] The figures are on loose sheets in the field book.

E

8

The Park farm and the Holkham sheep-shearings, 1776-1821

A FARM had been kept in hand at Holkham for many years before T. W. Coke came there, but he and his advisers enlarged it and made it famous throughout the world. The farming there was of the highest technical standard, and Coke's vivid personality and his forceful methods of advertisement attracted attention to his model farm.

By about 1780 the farm contained 1,300 acres or so of arable and marsh land plus over 200 acres of woodland.[1] For this, a notional rent of £875 a year was debited to the farm account. In 1788, 275 acres more land, taken from the Staithe and Honclecrondale farms, was added to the Park farm, additions for which 12*s*. or 14*s*. an acre were added to the rent, bringing it to over £1,000 a year. In 1790 the Longlands farm was taken into hand, probably including over 350 acres of arable and marsh. After 1808, 100 acres were taken in from what remained of the old Staithe farm.[2] By 1816, therefore, the farm in hand contained at least 2,000 acres, in addition to at least 200 acres of woods; since Coke came to Holkham it had become the largest farm on the estate and no expense was spared in stocking and equipping it. There were five barns on the farm when Coke came; soon he put up another, the Great Barn, 120 ft. long, surrounded by a yard with ample cattle-sheds, a building which, though it is now entirely mechanized, is still used for processing and storing the grain produced on the Park farm.[3] When two agriculturists, Ellman of Glynde and Thomas Boys, visited Holkham in 1792, they saw the 'Slaughter house, larder, &c. a very extensive and elegant building'.[4] At the end of 1816, livestock and implements on the farm were valued at about £5,000: there were

[1] G.E.D., 77. [2] A/B 1780-1809.
[3] A. Young, *Annals of Agriculture*, xix (1793), 451. There is a plan as frontispiece to the *Annals* (1804) and in R. N. Bacon, *The Report on the Agriculture of Norfolk* (London, 1844), opp. p. 395. [4] *Annals of Agriculture*, xix (1793), 114.

62 horses, colts, and foals, 160 head of neat cattle, 2,004 sheep, and 95 pigs.[5]

The importance of a farm like this in the hands of a rich landlord depended on its being managed in the best style, with room for experiments, and on the methods used on it being given the widest publicity, so that its example might encourage the spread of the best farming practices. What took place on the largest farm on the largest estate in Norfolk would have attracted attention, at least among Coke's tenants and neighbours, even if Coke had not taken steps to publicize his agricultural techniques. Coke produced an example of a farm cultivated in the best Norfolk manner, at which experimentation was continuous, and made sure that the results, sometimes, indeed, exaggerated or distorted, should be known everywhere. Agricultural visitors were welcome at Holkham. 'You could not please me more', Coke wrote to Arthur Young, 'than by recommending intelligent men to come to see my farm. It is from them I gain the little knowledge I have, and derive the satisfaction of communicating improvements amongst my tenantry.' In 1802 Coke urged Young, who was intending a Norfolk tour, to 'look upon my house as your own and favour me with as much of your company as your time will admit of.'[6] Many accounts of the farming there were obviously written under the influence of a cordial and hospitable reception. 'We meet here', noted Boys in 1792, 'with a reception far superior to that which arises from mere politeness, and which we plainly perceive will detain us longer than we intended.'[7] Boys was pleased to find that the best bull owned by Bakewell (who gave 'many uncivil answers to simple questions which we ask'd, [which] together with the form and ceremony of showing his stock was very disgusting') was inferior to one of Coke's.[8]

But the principal means of making known the farming methods used by Coke was the series of Holkham sheep-shearings. It is not certain when they began, though the first meeting is supposed to have been soon after Coke came to Holkham. Reports of the sheep-shearing of 1819 appeared under the heading 'Forty-third Anniversary'.[9] The last meeting, in 1821, was also described as the

[5] 'Farm Account, 1817-26' (Game Larder). The valuation excludes corn, etc., in store.
[6] B.M. Add. MS. 35127, f. 165, 23 July 1792; 35128, f. 470, 28 June 1802.
[7] *Annals*, xix (1793), 114.
[8] 5 Aug. 1792. T. Boys to T. W. Coke. Uncatalogued letter thanking Coke for his 'polite and Friendly reception' (Muniment Room).
[9] *Norwich Mercury*, 10 July 1819.

'Forty-third anniversary of that meeting'.[10] On the other hand, Young wrote in 1802 that 'this origin of sheep shearing meetings' had been held regularly 'for ten or twelve years'.[11] Probably the sheep-shearings became major events only after 1800 or so. The Norwich papers carried no reports of them before 1798, and in 1799 one of them called the duke of Bedford's sheep-shearing that year at Woburn 'the greatest meeting of the kind ever seen in England'.[12] Soon, however, Coke's meetings fully equalled Bedford's—many distinguished people went from one to the other, including Coke and Bedford themselves.

In 1820 Coke said that the meeting was first established for 'the improvement of the breed of sheep',[13] the contemplation of sheep was the main element in the meetings, and among the prizes awarded each year, the emphasis was on sheep-breeding. Here are the premiums offered for 1803:

> Cup, value 5 gns. for best two years old fat wether, in the wool, of the Leicester breed, bred in Norfolk.
> A similar cup for the best South Down two year wether, bred in Norfolk.
> Cup, value 10 gns. for the best Leicester theave and one, value 5 gns., for the second best Leicester theave, bred in Norfolk, and similar prizes for South Down theaves.
> Cup, value 5 gns. for the best boar.
> Five prizes of 1 to 5 gns. for the best shearers.
> Cup, value 10 gns. to the person in Norfolk who should produce the most satisfactory account of experiments that have been made on not less than ten acres of land to ascertain the comparative merits of the drill and broad-cast husbandry.
> Cup, value 10 gns. to the person in Norfolk who shall convert into water meadow the greatest tract of land, in proportion to the size of their farm, between 1 April 1802 and 1 June 1803.
> Cup, value 10 gns. for feeding the greatest number of horses with Swedish turnip: the quantity given, the expense, and the condition of the horses was to be reported.[14]

At each show, sheep were sold from the Park farm, some of Coke's

[10] R. N. Bacon, *A Report of the transactions at the Holkham Sheep-shearing . . . July 2, 3, 4 and 5* (Norwich, n.d. (?1821)).
[11] *Annals*, xxxix (1803), 61.
[12] *Norwich Mercury*, 9 June 1798 and 29 June 1799.
[13] Ibid., 8 July 1820.
[14] *Annals*, xxxix (1803), 64-6.

best rams hired out, and some cattle sold. In June 1806, for instance, these transactions took place at the sheep-shearing:[15]

	£	s.	d.
14 Leicester rams sold	270	8	0
32 Leicester theaves sold	223	13	0
58 Leicester ewes sold	399	0	0
Leicester lambs sold	184	16	0
86 Southdown theaves sold	463	1	0
1 Southdown ram sold	31	10	0
21 Southdown rams let	577	10	0
Devon cattle sold	85	1	0
	£2,234	19	0

Agricultural implements were shown each year: in 1803 for instance, there were Mr. Burrell's drill for sowing oil-cake dust, as fertilizer, with turnip seed or grain, and the Revd. Thomas Crowe Munning's improved drill. Every year the Park farm was looked over, and the farms of neighbouring tenants were visited. In 1820, for example, the company 'had the immense pleasure of inspecting the immense quantity of 70 acres of mangel wurzel' on Gibbs's farm at Quarles.[16] Reeve's water-meadows at Wighton were studied frequently.[17] High conviviality always prevailed. It reached its peak when the entire assembly dined with Coke in the afternoon of each of the three or four days of the meeting. These agrarian banquets were attended by larger and larger numbers every year. Two hundred dined in July 1804; 600 in July 1818.[18] Frequent toasts were drunk and speeches made. As an example, on the first day of sheep-shearing of 1810, toasts were drunk to 'Constitution and King', 'The best use of the Plough', 'Prosperity to Agriculture', 'Breeding in all its Branches', 'Small in size and great in value', 'A fine fleece and a fat carcase', 'Sir Joseph Banks', 'Mr. Tollett', 'Earl Fitzwilliam and Lord Milton', 'Mr. Coke' and 'The Duke of Bedford';[19] when Lord Bradford was toasted in 1819, 'His Lordship returned thanks and favoured the company with a song'.[20] The guests included men of great eminence, with a bias towards

[15] Holkham sheep-shearing account book, 1804-7 (Game Larder).
[16] *Norwich Mercury*, 8 July 1820.
[17] e.g. *Annals*, xxxix (1803), 61.
[18] *Norfolk Chronicle*, 5 July 1804; *Norwich Mercury*, 13 July 1818.
[19] *Norfolk Chronicle*, 30 June 1810.   [20] *Norwich Mercury*, 10 July 1819.

prominent Whigs and Radicals: in 1819 Robert Owen explained his plan for abolishing pauperism, and in 1821, Joseph Hume, who, later on, as we shall see, refused to allow the sharp edge of his zeal for reform to be blunted by memories of Coke's hospitality, was the principal speaker. In 1821 Francis Burdett and, as often, the Whiggish royal personage, the duke of Sussex, were also present. The American ambassador sometimes came, and there was nearly always a Russell or two. Other noblemen, and gentlemen who were neighbours or friends of Coke, visited Holkham for the sheep-shearings; there were agriculturists from all over the country and the numbers were made up by Coke's tenants and other Norfolk farmers.[21]

The sheep-shearings were an excellent device for advancing agriculture. Norfolk farmers could hear the theories of other counties; men from other counties could see the practice of Norfolk farmers. The after-dinner debates that frequently arose provided an informal agricultural forum. Coke's sales and hirings of stock, profitable to himself, must have been beneficial to local farmers' breeds. Those farmers were stimulated not only by the prizes, but also by the interest taken in their work by the glittering assemblage of the sheep-shearings. The tenants of farms nearby were probably both flattered and spurred forward by the visits to their farms of the Holkham guests. Indeed, the sheep-shearings probably contributed to the massive popularity Coke evidently enjoyed among his tenants. The writer of one of Coke's obituary notices remarked that 'he led the way in raising the character of our farmers and of our yeomen, and in elevating them to that station in the scale of society which they now so deservedly occupy'.[22] The sheep-shearings were a mark of the importance of Norfolk agriculture and of its principal exponents—the tenant farmers—meetings at which tenants mixed freely with their landlord and his eminent guests, whose praises flowed readily. In 1808 Coke toasted one of his tenants, 'Mr. Reeve, and may his example of good husbandry be followed throughout the United Kingdom'; such a toast before such a gathering must have given pleasure. The same Mr. Reeve in 1811 challenged some remarks made by his landlord on Devon cattle and asserted that he would produce home-bred stock equal to any Devonshire cattle in the kingdom. The next year he

[21] *Norwich Mercury*, 10 July 1819; 7 July 1821; *Norfolk Chronicle*, 27 June 1812, etc.
[22] *Norfolk Chronicle*, 9 July 1842.

produced, for a bet, a 'capital Norfolk home-bred three year old heifer' to rival a 'beautiful Devonshire fat heifer' of Coke's.[23] The sheep-shearings encouraged, then, what was so frequently toasted there, 'a good understanding between landlord and tenant'.

Much of the detail of the management of the home farm cannot be found. It does seem, though, that no major innovations can safely be claimed to have found their first origins there. Its importance was rather in accelerating the spread of the innovations of other farmers and breeders throughout Norfolk, in perfecting them, and in helping to spread the best Norfolk farming methods in the rest of the country. The well-managed home farm, amply supplied with capital and publicized through the sheep-shearings, was a good instrument for this purpose and is an outstanding example of what a rich and influential landlord could do. For example, though Coke stated, in 1821, that it was at Holkham that mangel-wurzel was first cultivated in Norfolk, Sir Mordaunt Martin, a neighbour of Coke's in the Burnhams, could have disputed it. But Sir Mordaunt admitted that he had

... not been at the expense of any accurate experiments; and if I had, should not expect to do much good by registering them, unless I should find less difficulty than I do in persuading my neighbours to step a few yards out of their road to see my practice in the culture and application of it. But I have the satisfaction of having the same good neighbour for my colleague in cultivating it, whose example has convinced this neighbourhood that Cooke's drill is a more valuable instrument than you [presumably Young] used to think it ever would prove. He this year ordered at least ten for his friends:

Coke could publicize new implements and new crops in a way Sir Mordaunt could not.[24]

The most important aspect of the management of the Park farm, the one which had the widest effects on other farms in Norfolk, was probably the various changes in the types of livestock kept there, most notably sheep, for the invasion of Norfolk farms by the Southdown breed was in great measure Coke's work.

The first of Coke's innovations in the type of sheep kept at Holkham, however, was the introduction of Bakewell's Leicesters. Coke told his guests, at the 1806 sheep-shearing, that many years before, finding his Norfolk sheep a 'vile degenerate breed', he had

[23] *Norfolk Chronicle*, 29 June 1811, and 27 June 1812.
[24] Sir Mordaunt Martin, Bart., 'Mangel-Wurzels', *Annals*, xxxii (1799), 273–5.

hired a Leicester tup from Bakewell, to improve them, and he had 'advised his tenants and the county to do the same'.[25] It is not certain when this happened, but he bought some before 1784[26] and new Leicesters were established at Holkham, though still side by side with pure-bred Norfolks, when Ellman (of Glynde) and Boys visited Holkham in 1792.[27] Coke then told them that he was about to sell all his Norfolks. It was Ellman himself, the improver of the Southdown breed, who persuaded Coke to buy Southdowns.[28] Ellman sent him some (from the flock of Ellman of Shoreham) in 1792 and more (from his own flock) in 1793.[29] Thereafter Coke kept Leicesters and Southdowns, comparing their virtues, as others were doing elsewhere.[30] Eventually he came out for Southdowns and in 1806 he said that 'after a fair trial of thirteen years' he found the Southdowns more profitable than the Leicesters. After 1806 he was selling his Leicester rams.[31] By 1811 prizes for Leicesters were discontinued at the sheep-shearings[32] and by 1817, at the latest, his flock of sheep was entirely made up of Southdowns.[33] A comparatively brief trial of Merino sheep failed to disturb their primacy. At the 1805 sheep-shearing, George Tollet showed two Merino rams, and on 20 August he wrote from Blackpool to Coke to say how pleased he was that Coke was experimenting with Merinos.[34] At the 1806 meeting, Coke declared that he would give Merinos and a cross of Merinos and Southdowns 'a fair trial for the benefit of the county at large'.[35] Prizes for Merinos were added to the list offered at the sheep-shearings. In 1808 Coke 'questioned whether the Merinos would not beat the South Downs'; but by 1811 he had given them up again.[36]

[25] *Norfolk Chronicle*, 28 June 1806.
[26] *Annals*, ii (1784), 378. Coke still preferred the mutton from the Norfolks, according to Young. [27] Ibid., xix (1793), 115.
[28] Coke at sheep-shearing of 1806, *Norfolk Chronicle*, 28 June 1806.
[29] *Annals*, xix (1793), 446. Arthur Young arranged the terms of the first sales: 80 ewes from Glynde at £1. 15s. 0d. each and 100 from Shoreham at £1. 11s. 6d. plus 40 ewe lambs at 16s. and 11s. respectively. B.M. Add. MS. 35127, f. 206.
[30] e.g. the duke of Bedford. *Norwich Mercury* on Woburn sheep-shearing, 29 June 1799.
[31] Holkham sheep-shearing account book, 1804–7 (Game Larder).
[32] *Norfolk Chronicle*, 29 June 1811.
[33] Stock Book 1814 (Game Larder). Yet he bought five Leicester rams from Smith of Dishley in 1820 at 20 guineas each. And in 1821 he recommended Southdowns crossed with Leicester (*Norwich Mercury*, 7 July 1821).
[34] Uncatalogued letter to T. W. Coke (Muniment Room). Coke bought six Spanish ewes for £24. 3s. 0d. each from Sir Joseph Banks. B.M. Nat. Hist. Botany Dept. Library. Banks Corresp. XVI, f. 120 copy Coke to Banks, 27 Aug. 1805.
[35] *Norfolk Chronicle*, 28 June 1806. [36] Ibid., 25 June 1808; 29 June 1811.

What was the effect outside the Holkham home farm of Coke's experiments with sheep? Coke cannot be credited with the introduction of Southdown sheep into Norfolk, or even into northwestern Norfolk. According to Young, Mr. Macro bought a flock of them for his Norfolk farm shortly before his death (in 1789 or 1790). When Mr. Macro died, the earl of Orford bought this flock and established it at Houghton—ten or eleven miles from Holkham.[37] On the other hand, their spread was certainly accelerated by Coke. His sheep-shearings were in great part campaigns against Norfolk sheep; each year prizes were given for other breeds bred in Norfolk, Southdown rams were let and ewes sold, and many of Coke's speeches to his guests were violent attacks on the Norfolks. In 1808 he declared that he

... had done everything in his power to extirpate the Norfolk sheep, the most worthless that could be kept, he had with great difficulty induced most of his tenants to change their stock, and they had found a great advantage in so doing; a few of them still retained their old prejudices; but this he would plainly tell them, that if they could afford to keep such an unprofitable breed of sheep upon their farms as the Norfolks were, it would fully justify him in raising their rents at the expiration of their leases.[38]

The farmers of Sussex sent Coke a gift in 1803, and John Ellman, the improver of Southdowns, wrote himself: 'I expect in the course of a few days you will receive a lot of live lambs, from the farmers of this County as a small token of their gratitude and respect for the liberal manner in which you have stepped forward in support of our Interest in introducing the Southdown breed of Sheep into Norfolk: An obligation I hope never to be forgotten in the mind of a Southdown farmer'.[39] Evidently they ranked Coke's efforts highly. We can be certain that they had a powerful effect on his tenants. General Fitzroy, the tenant of Kempstone farm, wrote to Blaikie in 1816: 'The stock on my Farm has been selected with great care Expense and attention, I changed the Leicester Flock for South Downs and the Home bred Cows for Devons. In short I have throughout had but one View which was to meet Mr. Cokes wishes in Everything.'[40]

The effect of Coke's favour for Southdowns outside his own

[37] *Annals*, xiii (1790), 162, and xv (1791), 305-6.
[38] *Norfolk Chronicle*, 25 June 1808.
[39] John Ellman, Glynde, to T. W. Coke, 2 Nov. 1803, uncatalogued (Muniment Room).
[40] 1816, L.B., 259.

estates is less certain, and can only be guessed at. The excellence of his flock was recognized: Arthur Young wrote in 1801, 'Mr. Coke's South Down flock at Holkham is so generally known to be excellent that I need only mention them. They are remarkably good, of the largest size, straight, fine woolled, clean.'[41] This example of the breed at its best was a good advertisement. Coke's hiring of rams and sale of ewes at the sheep-shearings provided a convenient opportunity for others to acquire or improve a Southdown flock. For although tenants were prominent at these sales and hirings, others took part: in 1806, for instance, twenty-one Southdown rams were let (for fees totalling £577. 10s. 0d.) and most of them were let to people who were not tenants.[42]

The other major change in Norfolk live-stock that Coke advocated was the introduction of North Devon cattle, a small breed, notable for the high quality of its beef. When Young wrote his *General View* of 1804, he reported that Coke had 'many Devons'[43] but he wrote of Devons in Norfolk as if they were of fairly recent introduction. By 1817 nearly all Coke's cattle were of this breed. Of the 160 head of neat cattle on the Holkham farm on 1 January 1817, 41 were Scots cattle, and all the rest Devons: dairy cows, beef stock, and working oxen.[44] In 1821 Coke summed up his conclusions: 'The Devon breed of cattle are much preferable upon poor land, to either of the two most celebrated breeds—the Shorthorns and Herefords. The Norfolk cattle are mongrel-bred animals, and many of them are the worst possible description.'[45] Coke encouraged the spread of the Devon breed by the same methods that he used for the Southdown sheep. Prizes were introduced for Devon cattle at the sheep-shearings some time soon after 1803, and prizes were given in 1811, for instance for the best Devon bull and the best two Devon heifers.[46] Devon cattle were offered for sale at the sheep-shearings, though in no very great number: £130 worth were sold in 1807, for example.[47] The new breed was frequently

[41] *Annals*, xxxvii (1801), 600.
[42] Holkham sheep-shearing account book, 1804-7 (Game Larder). It is striking that the fourteen Leicester rams sold that year fetched average prices of under £20 each while the Southdown rams were let at £27. 10s. 0d.
[43] A. Young, *General View of the Agriculture of the County of Norfolk* (1804), 446.
[44] 'Stock Book, 1814' (Game Larder).
[45] R. N. Bacon, *Report of the Transactions at the Holkham Sheep-shearing on Monday, Tuesday, Wednesday and Thursday, July 2, 3, 4 and 5* (Norwich, n.d. (?1821).
[46] *Norfolk Chronicle*, 29 June 1811.
[47] Holkham Sheep-shearing account book, 1804-7 (Game Larder).

advocated by Coke in his speeches, though he did not support them with the vehemence he used when he was calling for the spread of Southdown sheep. We have already noted that one tenant gave his adoption of Devons as an example of his deference to Coke's wishes. The Devon breed of cattle was probably never adopted in Norfolk, even on light land, as widely as was the Southdown breed of sheep.[48]

An innovation in technique encouraged by Coke both by precept and example was the drilling of seed. Until the later years of the eighteenth century seed was generally broadcast, in spite of Tull's work. There is, once again, no reason to imagine that Coke *introduced* row culture into north-western Norfolk, but he certainly helped to spread it. He was a consistent advocate of drilling, argued for it and practised it to good effect; Young found that the whole of Mr. Coke's 'great and highly improved farm' was 'well drilled' when he attended the sheep-shearing of 1802. 'Drilling never offered such an exhibition before', he wrote.[49] Implements on the farm at the end of 1816 included three double box-drill machines, three small barley drills, and one Northumberland drill, though there were also two broadcast sowing machines.[50] The effect of Coke's verbal advocacy can be seen at its most dramatic in 1819. At the sheep-shearing of that year, Sir John Sinclair, the president of the 'Board of Agriculture', spoke against drilling. Coke replied —in his usual emphatic way—and later in the meeting Sinclair announced his conversion and produced three resolutions he had just drafted expounding the virtues of drilling.[51] At the sheep-shearings prizes were offered to advance drilling: a prize for accounts of experiments in drilling has been noted already. Later on, in 1819, a prize was substituted for the best accounts of experiments to determine the comparative merits of drilling and dibbling seed.[52] Another prize, that for useful implements, was likely to encourage the showing of improved drills.[53] It is impossible to say how soon tenants began drilling. General Fitzroy of Kempstone farm wrote to Coke as late as August 1816 that part of one field on his farm was to be drilled and part broadcast. He was trying to convince his bailiff of the value of drilling.[54] Fitzroy, however, told

[48] *Report on the Agriculture of Norfolk* (London, 1844), 91, 301.
[49] *Annals*, xxxix (1803), 61.
[50] 'Farm Account, 1817-26' (Game Larder).
[51] *Norwich Mercury*, 10 July 1819.
[52] Ibid., 8 July 1820.   [53] A.L.B., i. 8.   [54] 1816, L.B., 126.

the sheep-shearing of 1819 that until then his bailiff had thwarted him in all his experiments with the drill.[55]

Those were three major innovations whose spread was encouraged by their exploitation on the Holkham farm. Further details of what was going on there are difficult to find and only a few facts emerge. Table J gives the acres of corn crops grown in 1782 to 1787.[56]

TABLE J

	1782	1783	1784	1785	1786	1787
Barley	219	174	268	140	213	159
Wheat	50	33	42	68	14	50
Oats	45	94	92	37	44	57
Rye	..	..	..	52	31	..
Peas	..	..	28	..	..	8
Beans	..	..	..	..	..	9

The preponderant importance of barley in the farm's economy—as elsewhere in that part of Norfolk—once more emerges clearly. The survival of rye growing is noteworthy, but clearly it is vestigial —possibly, indeed, it was not sown again on the Holkham farm, at least in Coke's time.

Yields per acre were recorded in coombs per acre (1 coomb = 4 bushels), as indicated in Table K.

TABLE K

	1782	1783	1784	1785	1786	1787
Barley	7½	8	8½	6½	7½	8½
Wheat	6½	3	4½-5	4½	7	6
Oats	..	9	10	5	11	11

These, unhappily, are the only reasonably reliable figures of crop yields surviving from the whole period covered in this study. A rough comparison may be made with twentieth-century yields at Holkham. In 1951 the barley yield was about 10 coombs to the

[55] *Norwich Mercury*, 10 July 1819.
[56] Corn Book, 1780–7 (Game Larder).

acre, in 1972 about 14·5. In 1951 the least productive field sown with wheat yielded 6·4 coombs to the acre, while the average was over 10. In 1972 the average for wheat was 10·6.[57]

Production of grain was recorded in lasts, coombs, and bushels (20 coombs = 1 last), as shown in Table L.

TABLE L

	Barley			Wheat			Oats			Rye		
	l.	c.	b.	l.	c.	b.	l.	c.	b.	l.	c.	b.
1782	74	12	0	16	0	0	19	18	1		..	
1783	70	8	1	4	14	3	39	9	2		..	
1784	105	1	3	8	8	2	42	8	2		..	
1785	44	15	1	15	7	2	9	8	0	9	1	0
1786	77	1	2	4	16	3	24	17	1	7	5	0
1787	66	1	1	14	9	3	30	9	2		..	

There are no later figures suitable for precise comparison with these. But an analysis of corn sold in 1817 and of stocks at the beginning and end of that year suggests that the harvest of 1817,[58] produced 107 lasts 3 coombs of barley and 56 lasts 5 coombs of wheat.[59] These figures do not point to any drastic change in productivity at Holkham since the 1780s, when the extra arable added to the farm is considered—though some arable had probably been devoted to plantations since the 1780s. On the other hand, they do imply that there took place the reasonable measure of progress that one would expect.

[57] Based on 1 coomb barley = 2 cwt., 1 coomb wheat = $2\frac{1}{4}$ cwt.
[58] Not an outstanding harvest according to T. Tooke, *A History of Prices 1793-1837* (London, 1838), ii. 19-21.
[59] 'Farm Account, 1817-26' (Game Larder).

9

General Finance, 1776-1822

MOST of Coke's income came from his Norfolk estates, but some came from outside Norfolk. He drew money from what was left of the out-estates Lord Leicester had owned: Kingsdown in Kent produced £847 in 1776; Portbury in Somerset, £1,211; Minster Lovell in Oxfordshire, £821, including wood sales; and Bevis Marks in London probably yielded £200. We have seen that Kingsdown, Portbury, and Bevis Marks were sold in 1785-7 to pay for the Warham estate in Norfolk. Minster Lovell survived until about 1812 when most of Coke's property there was sold to pay for Egmere; Coke kept—until 1824—only a small area of woodland there.[1] Coke also inherited from his father other properties outside Norfolk which had never been owned by Lord Leicester, the properties in Derbyshire and Lancashire, which had come to his father by Sir Edward Coke's will of 1725.[2] We have seen that the Derbyshire properties were devoted to the support of Thomas William Coke's younger brother, Edward Coke. It is possible that Edward was also given the rents of the Lancashire estates, though later on T. W. Coke drew a small income from them, and it may be that Coke drew the income from them, of at most £4,000 a year, until the sales of 1790-1804.[3] Certainly Coke retained formal ownership over both the Derbyshire and the Lancashire estates. Another estate outside Norfolk, that at Hillesden in Buckinghamshire, came to Coke from his mother, Elizabeth, in 1803. It was worth about £4,000 a year, but an annuity of £3,000 a year was paid from its rents to Mrs. Coke until she died in 1810.[4] Finally, there remained the tolls paid by shipping for the support of Dungeness lighthouse—tolls which yielded enough to provide a growing balance every year to be devoted to the support of Coke. In 1776 the lighthouse provided £2,619 net; in 1816 £6,042.[5]

[1] A/B 1776, 1785-7, 1812-13, 1823-5.
[2] 1824, L.B., 163-4.
[3] The income is guessed from the amount raised by the sales—about £110,000.
[4] A/B 1803-10.
[5] A/B 1776, 1816.

Apart from the lighthouse, Blaikie, the steward, was justified in telling a tenant in 1822 that it was erroneous to suppose that Coke had 'resources independent of the land'. 'It is utterly impossible', Blaikie declared, 'for Mr. Coke to support himself and his Family if his Tenants pay him no Rents.'[6] Thomas William Coke, even more than Lord Leicester, was dependent on agricultural rents. Only through the lighthouse did he share in the increasing commercial wealth of England; he never held a paid office, and he had no other sources of income.

In 1776 Coke's gross income from all sources, excluding Lancashire, was £18,242. In 1786 it was £23,760 and in 1796, £28,262. By 1806 gross income had increased to £38,017, and by 1816, it had reached £47,974.[7] Net income was affected by arrears of rents, by casual profits from the estates (notably sales of timber), by taxes, and by repairs and other expenses. After deducting estate expenses, we get the following figures for net income from all sources, excluding any income from Lancashire (from which county there may have come £2,000 to £3,000 a year more until the estates there were sold). The figures are of average net income, to the nearest £500, during the years included:

1776–85	£17,000
1786–95	£19,000
1796–1805	£24,000
1806–15	£29,500

Spending on repairs has already been examined. The effect of wartime taxation is masked in these figures by the addition of Hillesden to the estates. In the years 1807 to 1816, £53,824 was paid for Land Tax and Property Tax (i.e. Income Tax): a proportion of 13·2 per cent of gross rents for those years. In 1807 and 1808, when the incidence of war taxation was greatest, 14·8 per cent of gross rents was paid for taxes. For Norfolk alone, the proportions are higher: 13·9 per cent of gross rents in 1807–16 and 15·9 per cent of gross rents in 1807 and 1808, while proportions of gross income (i.e. after adding casual profits) from the Norfolk estates are 13·3 per cent paid in taxes in 1807–16 and 14·9 per cent in 1807 and 1808. These proportionate levels of direct taxation are high by pre-twentieth-century standards, especially when it is remembered that

[6] 1822, L.B., 15.
[7] After 1803, the figures were swollen by the Hillesden rents. It is possible that another £3,000 to £4,000 should be added for Lancashire.

the assessed taxes (indirect taxes mainly on houses and domestic establishments) became quite severe during the Napoleonic war, particularly in its later years. In 1807 assessed taxes at Holkham were at a level of £564. 7s. 0d. a year.[8] But even with the assessed taxes of the later war years, the proportion of income paid out in taxes by Coke at the height of the Napoleonic war does not reach the proportion paid by the Guardians a century earlier for the campaigns of Marlborough's time.[9]

Certain other payments had to be made out of Coke's net income from the estates before it could be applied to his maintenance. The wife of Lord Leicester's son, Lord Coke who died in 1753, survived until 1811: this was the notorious Lady Mary. She was entitled to £2,000 a year as her jointure from the estate formerly held by her father-in-law.[10] Coke's mother received £3,000 a year after she handed over the Hillesden estate in 1803 until she died in 1810.[11] From June 1820 Thomas William Coke the younger (Coke's nephew and heir presumptive until 1822) was given an annuity of £1,000 a year (to be raised to £2,500 when he married).[12] Still more serious was the cost of servicing the debt of nearly £100,000 that Coke faced when he came into his estates. Lord Leicester had left debts of £90,973. 18s. 10½d., of which £5,750 had been paid off in 1768. Another £3,000 had been allotted to repayment of debt by the first year's operation of the sinking fund Lord Leicester's will had called for. Therefore, £8,750 should be deducted from the debt on the settled estate. Wenman Coke, however, left further debts, beyond what could be paid for from his personalty, of £15,055. 8s. 5d.[13] The result was that Thomas William Coke had to meet a yearly interest charge of at least £4,000. In theory this burden should have fallen away as the £3,000 sinking fund took effect, but repayment of the debts on the settled estate ceased after 1789. After that year, the sinking-fund provision of Lord Leicester's will received only the token acknowledgment of an entry in the audit books and after 1809 the wishes of Lord Leicester were disregarded as well as disobeyed.[14] Even before 1789, they were not observed with full

---

[8] 'Holkham Household & Other Accounts', 1801-7 (Game Larder).
[9] See p. 3.
[10] H.F.D., 52, and Audit Books. For Lady Mary, see J. A. Home (ed.), *The Letters and Journals of Lady Mary Coke* (Edinburgh, 1889), and Horace Walpole's correspondence.
[11] A/B 1803-10.   [12] H.F.D., 112, 'General Statement . . .'.
[13] See above, pp. 62, 69.
[14] A/B accounts current. The entries, 1789-1809, read 'Sinking fund: - - -'.

General Finance, 1776–1822

rigour. For by then only £34,257 of Lord Leicester's debt had been repaid, and £5,750 of this was repaid before the sinking fund began. It seems likely, as we shall see below, that, in spite of such repayment as did take place, Coke's debts did not for long fall below a total of £100,000; no doubt, indeed, it was the futility of repaying debt at the rate of £3,000 a year with one hand and borrowing the same amount, or more, with the other that caused the sinking fund to be neglected.

At least £6,000 a year, therefore, must be deducted from the net income stated above to produce a figure for disposable income and in the years 1803–10, at least £9,000 a year should be taken away. When it is noted that Coke's household and personal expenses were estimated in 1822 at just under £16,000 a year, it is not surprising to find Coke's debt increasing.[15] Especially when one considers the capital sums he must have raised to pay daughters' portions, and the interest charges that borrowings for them would have involved, it would seem reasonable to assume that Coke had often to borrow to cover a deficiency of income compared with spending. This was a dangerous process, since increasing debt meant increasing interest charges and so exacerbated the disease that caused the increase in debt. Coke's debts, however, did not become disastrously large; their increase was largely offset by the steady increase in the value of the estates. Only in the 1820s when the increase in income from the estates temporarily ceased did the debts seem dangerous—and even so, they were dealt with without grave difficulty.

How great did the burden of debt become and why? Coke began with debts of about £97,000—assuming that any debts of his own contracted before the death of his father, who died when T. W. Coke was twenty-one, were balanced by his wife's portion (he married Jane Dutton in October 1775). By 1822 Coke's debts were about £230,000.[16] There is a gap of about £133,000 to be explained.

This increase in debt cannot be explained by purchases of land, though, as has already been calculated, £51,000 or more was laid out on buying new land up to 1816 and another £10,000 was laid out between 1816 and 1822.[17] This amount of £61,000 or so was

[15] H.F.D., 112. Statement showing the annual 'Income and Expenditure of Thos. Wm. Coke Esqr.'.
[16] H.F.D., 112. Statement showing the amounts of Debts, Charges, Legacies, Annuities, etc.
[17] 1827, L.B., 31–3.

more than balanced by the amount received for the Lancashire estates—£110,000 in 1790–1804;[18] indeed this balance of purchases and sales produces more capital to account for: another £49,000, which, added to the balance of debt, makes £182,000 somehow disposed of by Coke between 1776 and 1822. Furthermore, Coke received £6,419. 16s. 11d. after 1810 as his share of his mother's personal estate.[19] A final total of about £188,000 of capital has therefore to be explained away.

It is impossible to provide a confident explanation of how Coke spent all this money. The only certain capital drain on his finances in those years was the provision of portions for his daughters by his first marriage: he had three, Jane Elizabeth, who married, firstly, Viscount Andover in 1796 and, secondly, in 1806, Henry Digby; his second daughter was Ann Margaret, who married Thomas Anson, later Viscount Anson, in 1794; and his third was Elizabeth Wilhelmina, who married John Spencer-Stanhope in December 1822. Each of these three daughters evidently received portions of no less than £30,000, though the two eldest did not get the whole sum as soon as they married.[20] Thus £90,000 capital can be ascribed to daughters' portions; leaving another £98,000 to be accounted for—apart from anything Coke may have received when he made his second marriage, with Lady Anne Amelia Keppel, in February 1822.

Spending on politics helps to explain Coke's spending of capital. Coke, the largest landowner in Norfolk, naturally possessed great political influence in the county, but its exploitation could be very costly. He was directly or indirectly involved in four contests for the county representation of Norfolk—in 1784, 1802, 1806, and 1817. In 1784 the contest arose at the last minute and Coke withdrew after only nine days of canvassing. The contest of 1802 was also brief, though at that election voters were brought to the poll.

[18] 1823, L.B., 117.   [19] H.F.D., 97.
[20] H.F.D., 87, shows that Ann Margaret Coke got £20,000 when she married. £20,000 from the Lancashire sales went to Anson—presumably to pay for it (1823, L.B., 117). No doubt Jane Coke got the same. Very likely these two marriages largely contributed to bringing about the sales in Lancashire. A General Statement of the Property and Affairs of T.W.C. of 1822 (in H.F.D., 112) recites a deed of 18 and 19 June 1816 showing that Coke proposed to give £30,000 to Elizabeth Wilhelmina and £10,000 more each to Ann and Jane. In 1822, £22,000 is recorded as a debt to be contracted for Elizabeth Wilhelmina (H.F.D., 112—Statement showing the amount of debts . . .). Perhaps the other £8,000 was given before October 1822 when that note was made. It must be said that it is not certain that the girls eventually got more than £20,000 each.

The 1806 election in Norfolk was a full-scale battle involving vast expense. In 1817 Coke evidently gave financial support to Pratt, one of the candidates. Spending in 1784 was small: an account book at Holkham records 'a particular Account of the Disbursement on Election for Norfolk, April 1784' and contains 'the Expences attending the Canvass &c on the part of Sir Edward Astley Bart and Thos. Willm. Coke Esq . . .'.[21] The total is only £1,827. 19s. 4d. How much was spent on the 1802 election is not known though the opposition asserted, significantly, that the polling had been protracted by means of 'the overbearing weight of Mr. Coke's purse'. In 1806 the expenses of Coke and Windham came to nearly £33,000 of which about £11,000 was raised by subscription. Probably Coke paid out much more than half of the remaining £22,000. In 1817 everyone assumed that most of the money behind Pratt came from Coke. It can be confidently guessed that Coke's total spending on local politics must have been £40,000 at least, and probably more.[22]

As for the remaining capital disposed of, it can probably be ascribed to fairly consistent overspending; an average yearly balance of about £1,000 would account for it all.

Coke then, was a substantial borrower between 1776 and 1822, during which years his debts increased by about £133,000. Since some repayment of old debts took place in those years, the total new borrowing by Coke was higher still. After 1815 net repayment seems to have been going on, so that the burden of debt was built up before the end of the war. How and when were these moneys raised? What difficulties did Coke find in maintaining a high level of indebtedness? Unfortunately only partial answers can be given.

The ways in which Coke raised money are best shown by the list of his debts in 1822.[23]

	Principal	Interest
	£ s. d.	£ s. d.
Mortgages on the settled estate	56,716 18 10	2,835 16 8
do. on the private estate	45,000 0 0	2,250 0 0
Bonds	65,817 13 4	3,290 17 8
Promissory notes	20,740 0 0	1,037 0 0
Banker's balance, say	20,000 0 0	1,000 0 0
do. for Miss Coke	22,000 0 0	1,100 0 0

[21] Muniment Room.
[22] All the information in this paragraph, apart from that in the account book for 1784, has come from Brian Hayes, of Corpus Christi College, Cambridge.
[23] In H.F.D., 112. Statement showing the amount of debts, etc.

It will be seen that the interest on the debts, of all types, was then 5 per cent. The first item is what remained of Lord Leicester's debt from before 1759. It was not a very great burden for the settled estate to bear, for the capital value of that estate by then must have been somewhere between three-quarters of a million and a million pounds, at least. We have seen that Lord Leicester's will forbade any borrowing for any purpose on the security of the estate it settled. In consequence, Coke could offer as security for debts, even for raising portions, only those of his estates which were not in settlement. This fact may help to explain his defiance of the sinking fund provision of Lord Leicester's will—for, if he could not repay the debts on the settled estate without borrowing more money, he would be obliged to borrow on the estates not in settlement, thus concentrating more and more debt on the latter estates, and weakening his future borrowing power by doing so. Lord Leicester, of course, knew, when he made the will, that the future tenants-for-life of the estates he was settling would have the Lancashire and Derbyshire estates at their disposal.

There is no sign that Coke found any difficulty in delaying the repayment of the mortgages on the settled estate; it was apparently quite possible to avoid repaying them even at the time of the wars against the American colonists and against revolutionary and Napoleonic France, when yields on government securities rose to 5 per cent, the legal maximum rate of interest available on mortgages. Indeed Mrs. Tufton, whose mortgage for £8,000 was paid off in August 1778, under the sinking fund provision, insisted, when notice of repayment reached her, that 'as it was difficult to raise money at that time', Coke should not pay her unless it was 'perfectly convenient' for him to do so.[24]

The mortgages on the settled estates were sometimes assigned to new mortgagees, but their repayment never became unavoidable; difficulty in raising new loans because of shortage of money offered for lending to private individuals did not apparently make it difficult to keep old loans going. Some lenders, at least, were clearly willing to allow old mortgage loans to continue at times when they would not have given new ones; Coke might have been seriously inconvenienced if one of his creditors had called in his money at such a time. Coke's creditors did not always even exact as much

[24] Copy letter of T. W. C. to Ralph Cauldwell, 5 Sept. 1783, in bundle relating to *Coke* v. *Cauldwell* (Game Larder).

interest as they could have done. The interest rates on the big mortgage to Yorke and others—the one for £30,616—were the most sensitive to market changes. Rates (per cent) on this loan moved as follows:

1775	1776–7	1778	1779–92
4½ to Lady Day	4¼	4¼ to Lady Day	4½
4¼ to Michaelmas		4½ to Michaelmas	

1793	1794	1795	1796–1823
4½ to Lady Day	4	4 to 23 August	5
4 to Michaelmas		5 to Michaelmas	

By contrast the £10,000 due on mortgage to Henry Cavendish remained at 4 per cent from 1795 to 1809, when it was at last raised to 5 per cent.[25]

The mortgages on the 'Private estate' were all on the properties in Derbyshire. Though Coke allowed his brother to draw the rents from those estates, he kept for himself the power to offer them as security for loans. Apart from the Derbyshire estates there was little Coke could pledge except the lands he bought in Norfolk. The capital value of the Derbyshire estate cannot have been much above £135,000 to £150,000, so that it bore a much heavier proportionate weight of debt than the settled estate.[26]

Of the bond debts, £24,000 was due to Coke's brother: it is possible that this represented security for some family obligation assumed by Coke. The remaining bond debts were contracted during the later years of the war, in 1806, 1807, and 1811, and immediately after the war in 1817 and 1818. The most important lenders were Roger Wilbraham of Twickenham, and Thomas Sanctuary, a tenant, of Weasenham. Roger Wilbraham lent £32,000 in 1811 at 'legal interest', of which £18,000 was repaid in 1813; and he lent another £12,000 at 5 per cent in 1817. By 1816, £25,889. 19s. 10d. was due to Sanctuary.[27] The smallest bond debts were for £2,000 to each of two of Coke's daughters. There is no reason to suppose that lasting debts were contracted for smaller sums than that, even on promissory note, at least in the later years of the period 1776 to 1822. The only possible exceptions are £600 due to Coke's sister, Lady Hunloke, and £275 due to the 'Holkham Benefit Society', but it is unlikely that Coke was reduced to

[25] A/B accounts current.
[26] 1827, L.B., 31–3.
[27] Bonds in H.F.D., 92; 1827, L.B., 31–3.

requesting loans of any such body.[28] Thus, there is a contrast with the borrowings of Lord Leicester's time, in the early eighteenth century, when quite small sums were raised.[29] Probably the 'banker's balance' now fulfilled their function. A bank overdraft evidently contributed to keeping Thomas William Coke's finances afloat; there is no sign that an overdraft had come to Lord Leicester's help. In the later 1790s Coke's average overdraft at Gurney's bank seems to have been of the order of £5,000.[30]

In the years 1776 to 1822, then, a rising income did not prevent Coke's incurring more debt. After 1816, and more rapidly after 1822, total debts began to be reduced. The very important point can be made again for the period 1776 to 1822 that there is no sign whatever that financial difficulty for the landlord caused him to economize on investment in his land. Repairs to old farm buildings and the construction of new ones went forward on a lavish scale; the buying in of intermingled lands continued and enclosure advanced. Spending by the landlord on improvement of the soil itself was perhaps less than in Lord Leicester's time, but this was clearly not due to economizing on Coke's part but to a general movement among landlords in the area. In these years it is clear that Coke did not contribute any capital for direct or indirect investment outside his own estates; on the contrary, he drew in a substantial body of capital from outside them. So far as Coke was concerned, agricultural profits did not foster the advance of commerce and industry; if anything, it was the other way round.

[28] 1827, L.B., 31-3. £888. 17s. 6d. to 'Saffery, Downham' was almost certainly an obligation Coke had entered into to support a friend's credit.

[29] See above, pp. 31-4.

[30] Gurney MSS. (Barclays Bank Ltd., 54 Lombard St.). A 11/28 (6), f. 54, A 11/28 (7), f. 209, f. 309 record some half-yearly interest charges debited to Coke.

10

The estate under Coke, 1816-1842

i. *Francis Blaikie, Steward*

IN January 1816, soon after the end of the great war, a new agent came to Holkham: Francis Blaikie. Blaikie had been agent to Lord Chesterfield and was engaged by Coke when Chesterfield died.[1] His experience and abilities raised the prestige of the Holkham steward to heights never before attained and his power and influence to a point only reached before by Cauldwell, when he dominated the estate in Lady Leicester's time. Blaikie's eminence was reflected in his salary. Francis Crick, his immediate predecessor, had received, at the most, £300 a year and, earlier still, Samuel Brougham had apparently received as little as £120.[2] Blaikie started at £550, a salary which was raised within a year or two to £600, and then, by 1821, to a normal salary of £650. To assist him in the estate office Blaikie had a clerk, William Baker—who eventually succeeded Blaikie—at a salary of 100 guineas in 1818, soon raised to £155, and an 'architect', first Henry Savage, who normally received £100 a year, and later Stephen Emerson, who got £140 a year. Baker may have paid for an assistant for himself. When Blaikie retired in 1832, to be succeeded by Baker, the new steward received only £300 a year at first, though this was raised to £400 after a few years had passed.[3]

In return Blaikie provided a remarkable combination of zeal, energy, and knowledge. He was an agronome, the author of several treatises on agricultural matters: 'I consider it a duty incumbent upon man to diffuse to his fellow-men the fruits of his experience in this life. All men have not the same advantages of acquiring information. Those advantages are bestowed upon us by the providence of God.'[4] Blaikie sent some of his pamphlets to tenants, who

---

[1] R. N. Bacon, *Report on the Agriculture of Norfolk* (London, 1844), 346-7.
[2] A/B 1790, 1792, 1808, 1813, accounts current. Brougham's low salary may be explained by the calling in of outside advisers to manage part of the affairs of the estate.
[3] A/B, 1816, 1818, 1821, 1831, 1834, 1844, accounts current. [4] A.L.B., i. 78.

could thus share his advantages,[5] and with the deep sense of duty that was his leading characteristic, he arranged for further dissemination of his ideas through the simple device of asking questions in the *Farmers Journal* under 'various signatures and dated from various places' and answering them himself. This he did, he recorded, 'not infrequently'.[6] He was an inventor: in particular, he devised an inverted horse-hoe to clean between rows without smothering growing plants.[7] Blaikie was able and willing to give detailed suggestions, which amounted sometimes to instructions, to tenants on how they should use their land. In 1817, for example, he sent them a circular advising them against sowing damaged barley seed from the season before.[8] In May 1818 he wrote to a tenant who proposed, exceptionally, to follow a wheat crop by sowing barley with grass seeds in one of his enclosures, to recommend strongly that he 'strain every nerve to prepare manure compost of some sort, to top Dress the young layer as soon as the Barley crop is removed'.[9]

Blaikie's energy was soon evident. As soon as he arrived, he introduced the use of letter-books into which every letter coming into the estate office and going out from it was copied. He soon attacked Stokes, the local solicitor employed on estate business; in December 1816 Blaikie wrote briskly to him:

> I expect to be quite astonished at the multitude of official estate business completed—I assure you nothing gives me much greater pleasure than when I hear the word *Finished* pronounced, as applied to Legal and Official business . . . Now, if you please, *we two* will lay our heads together; shoulder to shoulder lift manfully and surprise the world, by a regular clearance of yours and this office, of the various and numerous unsettled Estate transactions, with which they are at present incumbered —We will verify the old adage of a long pull, a Strong pull &c. &c.[10]

He followed this up by an exceedingly impatient letter early in the New Year, and enclosed a list of business transactions he wanted concluded at once, promising a second list as soon as the first was disposed of: 'I trust you will excuse me should my zeal for regular dispatch of business lead me to express my sentiments on the subject with a little warmth.'[11] In 1828 the most articulate of the

---

[5] 1816, L.B., 265, 280; A.L.B., i. 181. [6] A.L.B., i. 77-8.
[7] 1816, L.B., 265; illustrated in R. N. Bacon, *Report*, 345, opposite which there is a good reproduction of the portrait in the estate office at Holkham.
[8] 1817, L.B., 59. [9] 1818, L.B., 115-16.
[10] 1816, L.B., 266. [11] 1817, L.B., 30.

The estate under Coke, 1816-1842

tenants, General Fitzroy, complained of delays in drawing up his lease and expressed surprise 'when I reflect on yr. acknowledged punctuality and dispatch of business'—Fitzroy was quickly snubbed.[12]

When Blaikie came to Holkham, Coke ordered him to 'take a general view of his Norfolk Estates and make a report to him'. Blaikie did this in the late summer of 1816 and arranged for the copying of his reports on the larger farms into a bound volume. His judgements on the tenants ranged from that on John Overman, 'a very deserving, industrious, attentive and persevering good Tenant . . . A Pattern Farmer', through, for instance, those on Charles Hill a 'young man' whose farm was 'a most disgusting spectacle' but whom 'experience may improve' to the brusque condemnation of the Widow Cooper's son 'who should be a comfort to her' as 'a profligate, drunken fellow'.[13]

Another, and important, mark of Blaikie's energy was his setting to work to draft a volume containing 'The General Form of Leases to be Granted by T.W.C. Esq. of his Estate in the County of Norfolk with Forms of Covenants for Cultivating the Arable Lands Settled and Approved by Francis Blaikie Gentleman in October 1817.'[14] Thenceforward all that was needed when a lease was made was the filling up of blanks in the new formulas.[15]

Blaikie's inexhaustible devotion to the interests of his master is evident on almost every page of the letter-books. He could never have provoked the suspicions aroused by Cauldwell; no one could have ventured even to dream of Blaikie's feathering his own nest at the expense of the estates. The massive series of his letters, magisterial compositions, untinged by any trace of humour—except for a blunt sarcasm directed against tenants who refused to 'condescend to be guided by the sound advice I give them in regard to the buildings on their farms' and insisted on erecting unnecessary buildings[16]—clearly demonstrate his conscientiousness. Tenants in arrears were made by Blaikie to feel that evasive conduct was sheer

[12] 1818, L.B., 89. In 1816 Fitzroy told Blaikie that he was 'well aware you speak your opinion manfully—without prejudice or partiality, and also I know you are a competent judge'. That is, of the management of a farm. 1816, L.B., 257.
[13] Blaikie's Reports, 3, 27, 14.
[14] Estate Office.
[15] Blaikie's use of the title 'Gentleman' is not without significance, and it contrasts with the insistence of tenant Reeve of Wighton, in 1816, that he should be addressed *'plain Mr.* not being entitled or *aspiring* to anything higher'. 1816, L.B., 280.
[16] 1827, L.B., 86.

ingratitude to their 'indulgent & Kind Landlord'—a favourite phrase of his.[17] There can be no doubt that Blaikie was utterly sincere when he wrote, in 1828, to Coke's London lawyer, that an increase in Coke's debts would be 'alike painful to your feelings and my own'.[18] It is not surprising to find Blaikie displaying, in 1822, paternal feelings towards one of Coke's daughters, and presenting her, as she wrote, with a calculation he had made of the income and charges of her betrothed.[19]

In September 1822 Coke's legal adviser, Philip Augustus Hanrott of Lincoln's Inn, wrote to Coke to suggest that Blaikie should be taken into their confidence and told everything about the crisis in Coke's affairs. 'I have no doubt that Mr. Blaikie's clear head & correct judgment will enable him to suggest many improvements in my plan—and therefore he should I think be in full possession of *all* my materials.' In particular Hanrott wished Coke to show Blaikie the draft of his will.[20] Thenceforward, until 1832, Blaikie and Hanrott directed all Coke's economic and financial affairs, subject to occasional difficulties in managing Coke himself.

When he retired in the summer of 1832, Blaikie was slightly at a loss. First he thought of visiting his brother in Scotland. Then he conceived a desire to study farming methods in Flanders, to go up the Rhine and to visit Switzerland to view Bonaparte's route across the Alps. Finally, he settled at Melrose, whence he wrote in 1840 a typically gloomy letter predicting future calamities for the agriculture of Britain from free trade and the 'awful encroachments of Democracy'.[21]

ii. *Leases and their enforcement*

Blaikie's code of leases was used, with a few modifications, for the remaining years of Coke's life. Its major provisions fixed the rotations tenants were to use under various systems applicable to various soils. First of all came the four-course rotation, Arable 'A'; this was the true Norfolk four-course:

1. Turnips for a crop, twice hoed and scoured, to be fed off and

---

[17] e.g. 1822, L.B., 53. [18] 1828, L.B., 17.
[19] Elizabeth Coke to John Spencer-Stanhope, in A. M. W. Stirling, *The Letter Bag of Lady Elizabeth Spencer-Stanhope* (1913), ii. 50.
[20] Hanrott to Coke, 3 Sept. 1822, in H.F.D., 112.
[21] 1832, L.B., 37-8; 1833, L.B., 21; A. M. W. Stirling, *Letter Bag*, ii. 168-9.

The estate under Coke, 1816-1842

consumed on the farm and as far as possible by sheep on the land on which they grow.
2. Barley laid down with grass seeds, not less than 12 lb. clover and 1 peck ryegrass per acre.
3. Grass of one years' layer mown only once or pastured by cattle or sheep.
4. Corn or grain.

The tenant was forbidden to sow oats in the last four years of his term except for his own horses. He was not to sow coleseed or rape-seed except to be fed off with cattle. He was not to grow more turnips than were needed for his farm. He was not to let any of his artificial grass stand for a crop of seed in the last three years of his term. He was forbidden to mow permanent pasture without the landlord's permission. He was not to mow his meadows more than once a year, nor two years in succession without improving by irrigation or top-dressing and manuring.

Then there was a five-course 'B':

1. Turnips as for 'A'.
2. Barley as for 'A'.
3. Grass of one year as for 'A'.
4. Grass of second year, not cut or mown, but pastured or fed by cattle or sheep or the land thoroughly summer-tilled.
5. White straw corn or grain or peas or beans.

There were added the same provisions as those appended to 'A'. If the tenant kept under grass for three years a piece of land sown with lucerne, sainfoin, cocksfoot, or other grasses, he might plough up an equal quantity after grass of only one years' lying to be followed by corn or peas or beans, but not more than two-fifths of the arable land at any time were to be under white straw crops. No successive crops of corn or grain were to be taken except after a three-year layer and then one of the crops must be peas or beans, so that two successive white straw crops might never be taken one after the other.

There followed 'C', an alternate four- and five-course: this was rotation 'A' followed by 'B'. Not more than four-ninths of the arable land should be sown and cropped with white straw crops in any one year.

Four six-course systems were included, numbered D1 to D4. D1 was:
1. Turnips
2. Barley
3. Grass
4. Grass
5. Peas or beans
6. Wheat, barley, or oats.

If sown grass failed after one year, the tenant might request permission to break it up and cultivate it as Coke or his agent should direct, but he was to bring it round to the six-course as soon as possible. Other provisions were as for the other rotations. The tenant might sow less than one-sixth of his arable with wheat or less than two-sixths with corn but not more. D2 was:
1. Turnips
2. Barley
3. Grass
4. Grass
5. Wheat, barley, or oats
6. Peas or beans.

Other provisions were as before.

D3 rotation, for 'very good strong soil', was:
1. Summerfallow, clear and complete
2. Barley
3. Grass
4. Wheat
5. Beans
6. Barley or oats.

This rotation was the only other one which postulated, like the four-course, 50 per cent white straw crops. It and the remaining rotation were the only ones not calling for turnips.

D4 for 'strong soil of second quality', was:
1. Summerfallow
2. Barley
3. Grass
4. Grass
5. Beans ('peas' added in pencil in margin)
6. Wheat, barley, or oats.

As Blaikie wrote about this time, the principal point was 'to convince a Farmer that it is not to his advantage, to take *quite* as many corn crops from the land he occupies as it will bear'.[22] Each set of covenants included an express and categorical provision against successive white straw crops. To this prohibition, and to the other provisions, was attached a penalty for default of £50 per acre per annum—an unenforceable clause useful only to frighten potentially disobedient tenants.

In 1824 a new clause was worked out to be added to all future four-course leases, applying to any time before the last four years of a tenant's term. If a tenant kept any of his arable in grass for two years, he might then take two successive crops of corn or grain from that land when the grass was broken up; but 'the meaning and intention of this course or order of Cropping is That in no case shall two White Straw Corn Crops be taken in succession—Nor more than one Crop of any sort of Corn or Grain immediately after a layer of one year only'. The effect was to give tenants under 'A' the option of cultivating their land under provisions 'D1' and 'D2' except in the last years of their term. It meant that no tenant was tied to the four-course, except for the last four years of his term.[23] Further refinements were introduced in the husbandry covenants after Blaikie retired. From 1833 tenants were forbidden to mow more grass for hay in the last year of their term than they had done in preceding years (leases must always be designed to frustrate an economical tenant's possible desire to leave the farm in a poorer state than he found it) and they were to deliver each year a schedule of their arable land stating what crops each field was bearing.[24] The latter clause one might expect in Blaikie's time, but it was Baker who reintroduced this check on tenants' conduct.

Blaikie's lease provisions included a clause directing the tenant to insure the farm buildings for a prescribed sum, clauses laying down the amount of cartage the tenant might be called on to do for his landlord, the usual provisions for the tenants to repair—with materials provided by the landlord—and an obligation to provide

[22] 1816, L.B., 129.
[23] 1824, L.B., 101. This new clause may explain the misleading statement of J. A. Venn, *Foundation of Agricultural Economics* (Cambridge, 1923), 35 that Coke's leases gave, for sixteen years 'complete freedom of action ... in regard to rotation, management of the land and sale of the produce; only in the last four years had the Norfolk four-course rotation to be restored'.
[24] 1833, L.B., 137; 1834, L.B., 90.

the landlord with a turkey for Christmas. Then there was an interesting and important clause laying down that the tenant 'shall and will effectually and substantially underdrain such part or parts of the demised lands as may stand in need thereof, or which shall be likely to be benefited thereby'. Underdraining was perhaps the most important nineteenth-century improvement (less important, it is true, on the light and dry soils of much of north-western Norfolk than in other parts of England). It is remarkable that tenants should have been expected to pay for it. This provision, together with the fact, already noted, that tenants in Coke of Norfolk's time normally paid for marling and claying, shows that improvement of the soil was much more the responsibility of the tenant and less that of the landlord than in the earlier years of this study.

What rotations did the leases granted after 1816 in fact call for? Rotations 'D3' and 'D4', for strong soils, were, it seems, never used at all; apparently Blaikie's covenants had been constructed to provide a sort of theoretical completeness.[25] For the rest, there is a marked preponderance of rotation 'A', the four-course. It was evidently normally used after 1816 and the exceptions became fewer as time went on. The exceptions involved rotations dictating a smaller proportion of corn-growing than the 50 per cent of the four-course. The usual rotation laid down when 'A' seemed unsuitable was 'C', the alternate four- and five-course, which limited white straw crops to four-ninths of the total arable acreage on a farm and called for one-third of the arable to be kept under sown grass. Naturally, this measure was applied when the land of the farm being let was inferior to land thought capable of bearing the four-course; the average rent per acre of land let under course 'C' was normally lower than that for land under course 'A'.[26] The use of rotation 'A', the four-course, or rather the permission to tenants to use the four-course (which was what that husbandry covenant

[25] It is odd that these rotations were never put in leases since there were some small areas of heavy land on the estate in Norfolk. Perhaps for some heavy-land farms leases were not granted at all; or perhaps none of Coke's farmers in Norfolk wished to accept rotations excluding turnips. 'Norfolk farmers are so wedded to turnips, that they sow them almost indiscriminately on all soils', wrote Arthur Young, *General View*, 219.

[26] 1820, L.B., 68; 1822, L.B., 13, 19, 30, 67, 74, 85, 87, 97-8, 122; 1823, L.B., 35, 46, 188; 1824, L.B., 101, 113; 1825, L.B., 67; 1827, L.B., 98, 105, 107, 109; 1828, L.B., 48, 71, 73; 1829, L.B., 36, 51; 1831, L.B., 31, 45, 64, 91, 98; 1833, L.B., 137-8; 1834, L.B., 65, 89, 90, 99; 1835, L.B., 43; 1837, L.B., 3, 20, 22, 78; 1838, L.B., 52, 57, 64-5, 82-3, 91-3, 111-12; 1839, L.B., 83; 1840, L.B., 16 each have one or more agreements for leases or instructions for drawing up the deed.

meant, modified as it was after 1824, by the option of extending the rotation into a six-course), spread further as farms which had hitherto been submitted to another regime came within its scope. In 1833, for instance, formal written sanction was given to a tenant, George Petingale, to manage part of his farm that had been tied to the five-course system on the four-course 'in consequence of the great improvement in the said land, made by the said George Petingale'.[27]

How far were the provisions of leases observed? This question is very difficult to answer. After Blaikie's advent, at least, administration at Holkham was generally thorough, and one would expect the maximum possible insistence on the husbandry covenants in leases. No doubt the steward cared little about what might be called beneficial infringements, such as the laying down of land to grass for a longer period than a lease required. His main endeavour would be to prevent excessive growing of corn crops, to prevent above half a tenant's arable being devoted to wheat, barley, or oats, and above all, to insist on the principle of the prohibition of successive white straw crops: 'The principle is uniform', declared Blaikie in 1821, 'and acted upon by all the Tenants upon the Estate whether upon Lease for a Term of Years, upon promise of a lease, or Tenants at Will holding from year to year.'[28]

It is certain that the provision in leases by which £50 an acre could be exacted for failure to keep to their terms was not enforceable. In the autumn of 1821, or the spring of 1822, a tenant, Gibbs, sowed with Talavera wheat about 30 acres which ought to have been sown with barley. The question was submitted to counsel whether Coke could claim £1,500 as a penalty, at the rate of £50 an acre. Mr. Serjeant Lens replied that £50 an acre could not be recovered but only such a sum as might be proved to be the damage.[29] There are no signs that Coke, or Blaikie, made any attempt in this or any other case to enforce the provisions of leases in the courts.

On the other hand, there are many instances of tenants seeking permission to deviate from the course laid down in their lease, a sign that they took the covenants seriously. Their requests were

[27] 1833, L.B., 7.
[28] 1821, L.B., 35-6.
[29] G.E.D., 112; 1822, L.B., 163. Relations with Gibbs were already strained. He was writing pamphlets, attacking Coke for failure to accept his nominee as his successor in his farm. It was perhaps only because of this that the question of legal action was raised. See E. H. Gibbs, *A Letter to Mr. Tuttell Moore* . . . (Holt, 1822), 10-11.

by no means certain to be granted, and when they were, the permission given was often hedged with conditions. In January 1817 Blaikie wrote to Isaac Rudd to permit him to sow oats out of course on a particular enclosure: 'this indulgence of sowing oats in the year 1817 is granted on the express condition that you clear out, summer fallow in a husbandlike manner and well dung the said piece of land in 1818—sow Barley and Clover seeds upon it in the year 1819.'[30] From Kempstone a tenant wrote in the same year for permission to sow 15 acres of oats on the wheat stubble—the land, he urged, was in high condition. Blaikie answered: 'So good a Tenant as you are, should have every reasonable indulgence granted—But were I to grant your present request, It would form a very dangerous precedent upon the Estate—I will give you a call the first favourable opportunity.'[31] When a tenant, Mr. Rix, wanted to sow barley after wheat (a common practice in the 1790s, as we have seen) in one field, he wrote very humbly, knowing it, as he said, to be 'a case to which both Mr. Coke and Yourself are averse'. Rix assured Blaikie that the field had been well clayed before the wheat had been sown and that he would manure it well when it came round again, in due course, to be sown with wheat. Blaikie gave a reluctant consent.[32] Often, however, a standard form of consent was given, as for instance: 'Memorandum Sept. 28, 1824. Mr. Blomfield of Warham to be allowed in the present year to Sow Wheat upon the Marsh breck 43 acres—out of the Common Course of Husbandry. The said field to be afterwards brought into the usual course of cropping. (Signed) Francis Blaikie.'[33] Blaikie's resistance to successive straw crops did not weaken as time went on: in 1831, he wrote to the executors of a tenant who wished to sow oats or barley on wheat stubble,

> I am aware that two white Straw Crops in Succession would not materially injure land of that quality, but the practice is contrary to the usages of Mr. Coke's Estate, and is very injurious to light land, and there is danger of an indulgence in one case, being drawn into, or quoted as a precedent in other cases, however, if the family consider it a material object to sow Barley or Oats on that piece of Land, I will, if they wish it, appeal to Mr. Coke for his decision on that point.[34]

The same policy was enforced when Baker succeeded Blaikie.

[30] 1817, L.B., 33.
[31] 1817, L.B., 86.
[32] 1818, L.B., 115–16.
[33] 1824, L.B., 134.
[34] 1831, L.B., 2–3.

R. N. Bacon wrote in 1844 as if lease covenants were taken quite seriously.[35] In 1850 Caird reported that farmers in north-western Norfolk were beginning to depart from the, by then, utterly orthodox, four-course, provided they were 'not restricted by rigorous covenants'.[36] It is hardly possible to doubt, on the evidence from which extracts have just been given, that the husbandry covenants of Coke's time were regarded by the tenants as obligations to be taken seriously. The documents quoted above show how an intelligent agent, backed by a rich landlord, could influence the conduct of tenant farmers.

The 21-year lease continued to be the normal contract on which Coke farms were let, though in the closing years of the war and the early years of the peace, there are signs that leases which fell in were not being immediately renewed—no doubt uncertainty about future prices induced hesitation. Thus, for example, when H. J. Overman took over a farm in Weasenham from the executors of T. Sanctuary at Michaelmas 1819, he held it from year to year until 1824; the rent up to Michaelmas 1821 was £849, but in the year ending Michaelmas 1822, it came down to £689, at which it remained. Blaikie's instructions for drawing up a lease to Overman were not issued until 1825; the term fixed was sixteen years to date from Michaelmas 1824.[37] Thus the lease would last until 1840, twenty-one years after Overman had taken the farm—the term was probably agreed on in 1819 at a time when it was felt that the rent could not be fixed for a long term of years to come. Again, the 21-year lease to Dewing of the Wicken farm in Castleacre expired in 1813. The rent had been £668. 5s. 0d. The farm was then let from year to year, at a rent of £1,100, until Michaelmas 1821. In January 1822 instructions were issued for a 21-year lease of the farm to be made out to Henry Abbott to date from 1821. The rent was to be £1,300 a year.[38]

The movements of rent of those examples illustrate the way in which a fall in prices, such as occurred after the war, might not, in the short run, reduce a landowner's income if his farms had been let on long lease in a period of rising prices or productivity. Under such conditions long leases protected a landlord against an abrupt fall in his income; a 21-year lease expiring in 1819, say, would have

---

[35] *Report*, 75-7.
[36] James Caird, *English Agriculture in 1850-51* (London, 1852), 171.
[37] A/B 1819-24; 1825, L.B., 86.   [38] A/B 1812-23; 1822, L.B., 19.

begun in 1798 and it would certainly be possible to raise the rent in 1819 even if the farmer's gross profit were shrinking. Furthermore, the high net profits such a farmer must have gathered before his lease fell in might provide him with reserves sufficient to tide him over a slump. The possibility for the landlord of raising some rents, even during a depression, when leases fell in, would counterbalance the necessity of reducing rents to other farmers who were in difficulties in consequence of paying a rent fixed in, say, 1811, 1812, or 1813. Even in 1822 some rents could be raised: the executors of Edmund Heagren gave up Crab Castle farm in Wighton at Michaelmas 1822, surrendering a lease for nineteen years from 1804. The rent had been £309. 6s. 6d. From Michaelmas 1822 William Wiffin took the farm on a 21-year lease at £400. The effects in the years of crisis after the war can be seen in the figures of rents shown in Table M. The arrears provide a very rough indication of the strains arising.

TABLE M

Figures for *settled* estate in Norfolk.[39]

	Gross rents due	Arrears outstanding
	£	£
1814	30,008	1,769
1815	30,883	3,168
1816	31,050	1,436
1817	31,038	1,202
1818	31,228	790
1819	31,651	1,571
1820	31,949	4,365
1821	31,271	7,585
1822	28,743	2,894
1823	30,657	2,089
1824	30,951	1,598

The stability shown by the figures for rents would hardly have been possible if rents had not generally been fixed under long leases. Landlords whose rents were on a year-to-year basis must have faced much more drastic falls in their income. Coke probably

[39] Rents were due at Michaelmas each year. I have deducted the allowances made to tenants in the years 1821-2 from the rents due. Arrears were rents unpaid when the books were closed—usually some months after Michaelmas. The low total of 1822 arrears is probably partly explained by the number of tenants who settled their rents after selling up their stock. Figures from A/B, to the nearest pound.

benefited, too, from the fact often asserted by Blaikie, that his farms were invariably let at very moderate rents. At the beginning of 1823 Blaikie claimed that 'it is well known that if many landlords were to reduce their rents 50 per Cent, their land would not then let so low (all circumstances considered) as a great part of Mr. Coke's Estate'.[40] For both reasons many of Coke's tenants could meet a depression comparatively easily. No doubt the farmers met difficulties at the end of the war, but there are few signs of severe stress on the estate before 1821.

iii. *The crisis, 1821–2*

Early in 1821 Blaikie saw that a crisis of extreme severity was approaching and he prepared to meet it. In February he worked out a series of questions the answers to which would show him how far tenderness to stricken tenants could be reconciled with the prior duty he owed to maintain his master's rights, and sent his questions to Hanrott, after showing them to Coke, for him to seek counsel's opinion. Serjeant Lens was applied to and he gave signed answers on 23 February 1821. Blaikie told Hanrott that it was 'probable in these most distressing times, that I may (I know not how soon) be unavoidably called upon, in the discharge of my official duty, to act upon principles unprecedented upon Mr. Coke's Estates'. His questions were about the circumstances in which failure to insist on payment of rent, when due, might lead to a weakening of the landlord's ultimate right to that rent. In reply, Lens explained the processes by which accumulated arrears of rent might be recovered; that a landlord could distrain on a tenant, even if the tenant had granted a bill of sale to creditors other than the landlord or assigned his goods for the benefit of such creditors; that a notice to quit to a tenant would not prevent distraint for arrears, whether the tenancy was from year to year or for a longer term; that the tenant's arrears in paying tithe, taxes, rates, or wages would not affect his liability to pay rent or arrears of rent.[41] The nature of the questions and answers is shown clearly enough in the last two:

Q. State the cases, wherein the landlord's right of precedence ceases and he becomes a Simple Creditor.	A. It is not usual to answer abstract questions as it may mislead, besides the difficulty of laying down general rules, without

[40] 1823, L.B., 10–11; A.L.B., i. 174. [41] 1821, L.B., 18–20.

	regard to the particular circumstances of a case. There is no such thing in Law as a right of precedence in a Landlord, except in the Case of Execution levied by the Sheriff before referred to [when one year's rent if due, was reserved out of the proceeds of the sale]. The landlord must in fact pursue some of the Remedies, with which he is by law furnished.
Q. If there are any other contingent cases, wherein the landlord would be entitled to, and justified in securing Himself, it is requested they may be pointed out, so as to give the Landlord a safe opportunity of extending his indulgence to His Tenantry.	A. ... it is not in my power to answer in an opinion a question so extensive. The Landlord can have no security beyond the visible stock on the premises and the general solvency of the Tenant.

The normal transactions of the sheep-shearing of 1821—the last meeting—were interrupted by cries of alarm. Coke propounded at length his panacea—the reduction of public spending and taxation—and he argued forcefully that only the selfishness, folly, and extravagance of ministers prevented the restoration of prosperity to farmers.[42] (Blaikie agreed about taxation but was also anxious that the 'circulating medium' should be increased.)[43] In spite of this dismay, not a single farmer on the estate gave up his farm in 1821, though only, no doubt, because the farmers were allowed to leave large sums in arrear, as many had already done in 1820. Total arrears in Norfolk (for the 'Private estate' as well as the settled estate) were over £7,500 when the books were closed for the year ending Michaelmas 1820, an increase of about £4,800 over the 1819 figures; when the books for the year ending Michaelmas 1821 were closed, arrears were about £10,500, an increase of about £3,000 that took place in spite of the rebates made to tenants that year of about £1,450. A year later the total of arrears had been reduced to about £5,500, after another £3,000 had been allowed to the tenants. The worst year was 1822 and the allowances to tenants do not explain this reduction in arrears; presumably, in

[42] *Norwich Mercury*, 7 July 1821.
[43] F. Blaikie to P. A. Hanrott, 24 June 1823 in H.F.D., 112.

The estate under Coke, 1816-1842

fact, many of the rents were settled by tenants who were giving up their farms and realizing their capital—a disastrous event for them.[44]

At the beginning of January 1822, we find Blaikie writing to a tenant, Mr. Purdy, who owed Coke £960. 3s. 2d., having paid no rent at all for 1821, and still owing £315. 15s. 10d. of his rent for 1820. Blaikie was doubtless spurred on by the knowledge that Coke's finances were weak, and he pointed out that 'Mr. Coke cannot even exist if He is deprived of the means' and concluded 'surely you can pay something'.[45] It seems that Purdy was constrained to give up his farm, which was taken over by John Hudson in the autumn of 1822.[46] Soon afterwards, Blaikie noted 'we have not yet seen the worst'.[47] He was right; the letter-book for 1822 is full of stories of despairing tenants. By June, Blaikie had almost lost hope: 'The whole landed interest, whether owners or occupiers are now nearly upon a par and their total destruction appears inevitable.'[48]

Reduction in Rents has no effect in getting the arrears paid up— Several farmers have resigned their leases—many others are on the point of doing so—I have let some of those farms at reduced rents with a proviso of still further reductions if times do not improve . . . In fact, it is utterly impossible (in most cases) that the occupier can make any rent from land at the present prices of produce; Labour, Poors rate and Taxes consume the whole. When there is a surplus, the Parson takes it, and the Landlord goes without and has to maintain the buildings into the bargain.[49]

Shortly afterwards a tenant wrote that he was selling 'as good wheat as I ever expect to grow' at 16s. per coomb and that his last crop of barley averaged 7s. 3d. [32s. and 14s. 6d. respectively per quarter] and he added: 'The misery and anxiety to carry on business in times like the present beggars all description.'[50]

The result was a great turnover in farms; until that year, as Blaikie told Hanrott in July, 'Mr. Coke has been induced from his innate goodness of heart, to continue the occupation of his Estate in the descendants of particular families; and it was a rare instance

---

[44] A/B 1820-3. It must be remembered that arrears recorded may conceal great delays in payments which were not entered as arrears. 'Arrears' were amounts due when the books were closed—the account to Michaelmas 1823, for instance, was not closed until the end of May 1824 (1824, L.B., 75). Fluctuations from year to year in arrears may only be due to fluctuations in the date of closing the books: figures of rent arrears should be used with great caution.
[45] 1822, L.B., 15.   [46] Ibid., 122. Of course, Purdy may merely have died.
[47] Ibid., 28.   [48] Ibid., 86.
[49] Francis Blaikie to P. A. Hanrott, 24 June 1822, in H.F.D., 112.   [50] 1822, L.B., 94.

that a stranger had an opportunity of obtaining an occupation under Mr. Coke.' This principle had to be abandoned: 'The working of Lord Castlereagh's wants have now effectually dissolved that social compact, and Mr. Coke has been under the painful necessity of throwing open his farms to public competition.' Even in 1822, however, and this is a striking fact, all the farms given up were relet to new tenants, and to what Blaikie called 'good men' that is 'possessed of good capital and the right sort of plodding industrious farmers'. Blaikie put this down to Coke's reputation as a landlord: 'A good name is better than great riches, so said Solomon, and his wise saying is verified in this case for while many other Landlords cannot find Tenants at any rent Mr. Coke has abundance of applications in anticipation of Farms dropping.' (Indeed, in November Blaikie was able to write to an applicant for a farm, 'There is not *at present*, nor is there likely to be *soon* a single farm upon Mr. Coke's Estates in this County either disengaged or unoccupied'.)[51] But, asked Blaikie rhetorically,

... will the New Tenants, (at reduced rents) be enabled to fulfil their contracts, and stand their ground, at the present prices of produce with the same weight of taxation, I give a decided opinion that they will not, indeed it is utterly impossible that they should do. No, the same painfully distressing scene will be acted over again until the country is brought to an indiscriminate mass of ruin.[52]

In fact, after 1822 recovery began.

Indeed, 1822 itself brought less damage to the Coke estates than might have been expected. Many tenants survived—some without any rent abatements. When General Fitzroy complained, at the beginning of 1823, at only being allowed 15 per cent rebate on his rent to Michaelmas 1822, and urged that other landlords were giving far more, Blaikie was able to make a powerful rejoinder, 'You, Honble. Sir, will probably be not a little surprised when I inform you that a great many (not less than thirty) of Mr. Coke's tenants paid their Rents in full at the last Audit and had no return made; and what is more, none asked for because none could conscientiously be required.' Blaikie admitted, on the other hand, that some tenants had had 30, 35, or 40 per cent of their year's rent remitted.[53]

The degree of difficulty undergone by tenants depended largely on how high their rent was and that depended on when their rent had been fixed. If they were holding their farm for the last years

[51] 1822, L.B., 157.   [52] Ibid., 100-1.   [53] 1823, L.B., 10-11.

The estate under Coke, 1816-1842

of a long lease, they might survive relatively easily. A rent fixed in 1801, 1802, or 1803 would be less onerous than one fixed in, say 1813. Thus the rent John Turner of Castleacre was paying for his farm in the difficult years, 1814 to 1821, was £920 a year. Before 1814 his farm had been on a 21-year lease, from 1793, at only £390. 11s. 4d. In 1822 Turner was allowed a rebate of £300, but this did not prevent him from throwing up his farm, which was let to another tenant at £700 a year.[54] A similar case emerged in 1824 when a tenant named Shipp was pressed for £107. 0s. 8d. arrears of rent—Shipp declared that he could not pay and burst out

> Not less (these last 11 years) have I sunk than all the *little* I had accumulated by incessant care and industry and for I then hoped my Old Age and Infirmity, but alas it is all vanished, and that by an honest answering an extravagant Rent as long as the means could be found— Nearly a double Rent was made on my occupation just as Agricultural produce lowered beneath a price to let Farmers live who had them at their Old Rents. How then, Sir, should it be otherwise with me than what has occurred, a transfer of my 31 yrs. labour care and attention to Holkham.[55]

Shipp had an understandable grievance for he held his farm on a 21-year lease from 1813—the very worst year for a tenant to have his rent fixed—at £362. 17s. 0d. a year. Before 1813 the farm had been at £200 a year.[56] Blaikie replied firmly to Shipp insisting that his rent was, in fact, moderate, that Coke had spent £1,607. 13s. 9d. in the past seven years on building and repairs on Shipp's farm and that Coke had 'a family dependent upon him'. If Shipp wished to throw up his lease, Coke would not object, but

> Mr. Coke has no wish to disturb you in the possession providing you pay the Rent regular, and maintain the farm in good husbandry state. To convince you that Mr. Coke does sincerely feel for you in your present painful situation, I have his authority to inform you, that if you are desirous of giving up your farm to him at Michs. next—He will, on your balancing your arrears of Rent, agree to pay you an annuity of £100 for the remainder of your life.[57]

Shipp asked to be allowed to pay after the 1824 harvest, and he kept his farm.[58] Such generosity and the courtesy of the following

[54] A/B 1812-22; 1822, L.B., 74, 98. [55] 1824, L.B., 75, 96.
[56] A/B Flitcham. [57] 1824, L.B., 98-9.
[58] Ibid., 114; A/B. In 1816 Blaikie had reported that Shipp was 'a most industrious and intelligent farmer and has done wonders in the improvement of his farm'. Blaikie's Reports, 11.

extract help to explain the devotion Coke inspired among his tenants: in the summer of 1822, Blaikie wrote to Shepherd of Wighton, 'I am directed by Him to say that he will not press you for the payment of any money at the ensuing Audit. He hopes that you will nevertheless make a point of attending to join the company at the dinner table—I again repeat what I have before told you, It is far from Mr. Coke's wish you should quit the farm if there is a possibility of your holding it to do yourself and your family any good. . . .'[59]

By contrast, as an example of a farm let at a rent fixed several years before the crisis, the Abbey farm at Castleacre was let for eighteen years from 1804 at £298. 7s. 0d. (probably that rent was fixed, in fact, in 1801, at the beginning of a 21-year period); previously, the tenant had held a 20-year lease, from 1781, at £200 a year. The tenant was not given any rent rebate during the slump, and the rent of the farm was not reduced when a new 21-year lease was agreed on in March 1823, to date from Michaelmas 1822, at £302 per annum.[60] It was even possible, sometimes, to raise the rent of a farm during the crisis: as we have seen, when the executors of Edmund Heagren gave up the Crabs Castle farm in Wighton at Michaelmas 1822 after it had been held on a 19-year lease from 1804 at £309. 6s. 6d. yearly rent (a rent probably fixed in 1802), William Wiffin agreed, in 1822, to take it for twenty-one years from Michaelmas at £400 a year.[61]

iv. Investment in the estates and the coming of 'high farming'

In 1823 recovery began. Rents steadied and even started to rise again, though it was not until 1833 that they got back to the level of 1820. Thereafter the rise went on: in the nine years from 1833 to 1842, the year of Coke's death, rents on the settled estates (whose acreage remained constant in these years) increased by 12 per cent. After 1836 the upward movement was greatly accelerated and in the five years from Michaelmas 1837 to Michaelmas 1842, rents rose by nearly 10 per cent. Table N gives the figures, to the nearest pound, for gross rents on the *settled* estates due at Michaelmas each year.

[59] 1822, L.B., 91.
[60] A/B; 1823, L.B., 46.
[61] A/B 1822; 1822, L.B., 85.

TABLE N

	£		£
1820	31,949	1832	31,862
1821	32,187 (of which £916 allowed to tenants)	1833	32,059
1822	31,680 (£2,937 allowed to tenants)	1834	32,266
1823	30,991 (£334 allowed to tenants)	1835	32,478
1824	30,951	1836	32,528
1825	31,092	1837	32,624
1826	31,135	1838	33,168
1827	31,297	1839	34,177
1828	31,470	1840	35,001
1829	31,657	1841	35,299
1830	31,790	1842	35,806
1831	31,790		

The crisis of the 1830s had no serious repercussions on the estate, though it was not until after it was past that rents began to rise rapidly; in 1834, it is true, arrears rose to £1,516, a level unusual by then, but trivial compared with the £7,585 of 1821. Otherwise little sign of distress—on the part of tenants, that is—can be found in the estate correspondence. In 1836 Coke was asked to sign a requisition, got up by the West Norfolk Agricultural Society, for a county meeting, to petition parliament to set up a committee to inquire into the distress of agriculture. His reply was in striking contrast to his impassioned denunciations of the government in 1821: 'As I am far from believing that it is in the power of the Government beyond what they have done, and will endeavour to do, to relieve the Agricultural Distress, I must beg leave to decline (though as Warm a Friend to that Noble Science as any member on the Committee) adding my name to a requisition to call a County Meeting at Lynn.' 'Supply and Demand', he added philosophically, 'will always regulate the Market.'[62]

After the crisis of the early 1820s, a slight slackening appeared in the rate of investment in the estates. The crisis caused Coke's advisers grave concern about the state of his finances, and it is possible, though not at all certain, that this led to more care being taken before extensive repairs or new buildings were approved. In any case, the fall in investment was not severe: in the years 1817–22, an average of £5,436 was returned from Norfolk rents for 'repairs', a term which still included new building. In the years 1823–32 the figure was £4,651, and in 1833–42 it was £4,998. As we have already seen, an uncertain amount must be added for materials

---

[62] A.L.B., i. 215. Coke's Whig friends were now in control of the government.

provided by the Holkham office, so that it would be unwise to put too much weight on these figures. These sums bear the following proportions to gross rents: 15½ per cent in 1817–22; 14 per cent in 1823–32; 14 per cent in 1833–42.

Coke himself took pride in the amount he had invested in his estates. In his will, he declared that he 'had expended out of his own income and property not less than Five hundred thousand pounds in various improvements . . . besides paying off and discharging all the incumbrances which were thereon [on the estates] when he came into possession'.[63]

Baker, the steward, commented after Coke's death, 'In respect to the £500,000 expenditure of the late Earl[64] that was entirely on the building department and included all materials whether paid for or growing on the Estate. It appears a very large sum but when it is considered the Number of Years the Estate was held and the immense buildings which have been created the average will not be thought much beyond what might be expected.' Baker added the next day: 'the £500,000 I have always considered fallacious in consequence of the Materials charged from this place being calculated at two thirds of the actual payments for workmanship, and added as such, but that being the case of course it cannot be altered and if it could I do not see that it would tend to any good.' Still, even if £500,000 was an exaggeration, it is clear that the figures for repairs and improvements, given above, once again very substantially understate the true cost. Sixteen years before, Blaikie had made an important and sometimes neglected economic point when he had complained

... that Mr. Coke's tenants *Generally* are much in the habit of erecting unnecessary buildings, and frequently do so without due consideration, such buildings are not only attended with uncalled for expense to the Landlord in the first instance, but entail a lasting encumbrance upon his Estate. For every particle of building not absolutely wanted is an encumbrance to the Estate, and a deterioration to the property. These remarks apply more immediately to Mr. Coke's Estate than any other in the Kingdom.[65]

---

[63] H.F.D., 131, copy of will of February 1838. Coke, of course, was under obligation to pay off those incumbrances, by the terms of the will of Thomas, earl of Leicester.

[64] Coke was created earl of Leicester in 1837.

[65] 1843, L.B., 139, 140; 1827, L.B., 86. As Baker pointed out, there would have been nothing to stop Coke from using the £500,000 for 'Horse racing or any other Gambling transaction if he had been so inclined'. 1843, L.B., 147–8.

The estate under Coke, 1816-1842

The high standard of construction and repair of farm buildings on Coke's estate evidently helped to induce other Norfolk landlords to keep their buildings in good condition. A witness was asked by the House of Commons Select Committee of 1833 whether farm buildings in Norfolk had deteriorated since 1825. 'No', he replied, 'they are better, because Mr. Coke bears so large a proportion of the whole and all his are in such excellent condition.' He went on to explain the good condition of buildings on other properties: 'Mr. Coke has partly caused it by the example he has set.'[66]

Coke's investment in his existing estates, as Baker's remarks confirm, was in farm buildings rather than in improvements to the soil, and in Coke's time capital needed for soil improvement was provided by the tenants. They were protected from loss by long leases—or by their confidence that they would not be unreasonably ejected from a farm—and sometimes also by the possibility of payment from incoming tenants for unexhausted improvements made by outgoing tenants. Thus a tenant in Weasenham who gave up his farm in 1834, was paid by the incoming tenant for 16 acres of land he had clayed, and he had paid the tenant before him for claying in a similar way.[67] Tenants were expected to marl 'such part, or parts of the land as require that application ... a quantity of marl to be laid on as may be considered necessary and beneficial according to the rules and practice of good Husbandry' as a prospective tenant from the Isle of Wight was told in 1829.[68] General Fitzroy of Kempstone wrote in 1816 that he regularly clayed 'from 20 to 30 Acres Annually according to circumstances'.[69] We have already seen that tenants were responsible for underdraining.[70] Tenants needed very substantial capital. Yet there seems never to

[66] Evidence of R. Wright, land agent, *Report from the Select Committee on Agriculture*, Parl. Papers 1833 (H.C. 612), V, p. 113, qs. 2095-7.

[67] 1834, L.B., 107.

[68] 1829, L.B., 38. He was told, too, that the greater part of the substantial Harpley Dam farm in Flitcham, which he was considering, 'requires marl'.

[69] 1816, L.B., 257.

[70] In 1829 General Fitzroy's manager wanted to buy from Holkham some tiles for drainage (1829, L.B., 34). West Norfolk draining habits and management of strong lands left something to be desired. Blaikie wrote in 1828 'A Norfolk farmer thinks nothing of expending £3 or £4 an Acre every fourth year in cleaning and manuring his fallows. While he boggles at expending £1 an Acre in the permanent improvement of under-draining strong land. Improvements on Arable land *may* pay £10 p.Cent on Capital, *under particularly favourable circumstances*. Improvements on Grass land, especially draining wet land, *may* and generally *will* pay from £50 to £100 p.Cent on Capital expended when the work is done judiciously' (ALB, i. 101).

have been any shortage of tenants in the first half of the nineteenth century. In 1831 Blaikie wrote to the friend of a prospective tenant: 'You very justly infer that Quarles farm is not to let at present, nor is there any other farm upon the Estate disengaged. If you will favour me with your friend's name ... I will make that entry in the list of candidates for farms.'[71]

After 1816 investment by acquiring more land continued, and consolidating purchases were made almost every year. From 1816 to 1839 slightly more than £24,000 was laid out on buying small properties. The largest single purchase was made in 1820, for £3,586. Many of the purchases were of small allotments of the heath at Wells, which had been enclosed during the war.[72] On the other hand, no substantial self-contained estate was bought. Indeed in 1827 Coke declined Lord Spencer's offer to sell him the substantial neighbouring property of North Creake because of caution induced by his financial problems in the 1820s.[73]

The years from the end of the war to the death of Coke in 1842 were years of advance towards increasingly intensive use of land, of the adoption of the practices grouped together in the vague phrase 'high farming'. In north-west Norfolk 'high farming' involved established eighteenth-century practice—a rotation of roots, short leys, and corn, with a large arable sheep flock—with more and more widespread use of the four-course rotation plus more and more extensive application of fertilizers. High farming was based at its peak on cheap labour, high prices for farm produce and new artificial fertilizers. Until chemical fertilizers became prominent, a well-managed four-course was the most effective form of 'high farming' in north-west Norfolk. With the appearance of increasingly cheap imported or chemical fertilizers, it became possible to modify the four-course, notably by growing more corn.

We have seen that the four-course was rare on the Coke estates in the 1790s. Evidently it spread rapidly in the first two decades of the nineteenth century, and continued to spread more slowly thereafter, until it was almost universal in the 1850s. The somewhat tentative nature of its use early in the nineteenth century is shown by a letter from General Fitzroy to Blaikie in which he explained that 'from the proportion of Grass land I have in proportion to

[71] 1831, L.B., 28.
[72] 1827, L.B., 31–3 for purchases 1816–26 and A/B accounts current for 1827–39.
[73] See p. 195.

The estate under Coke, 1816-1842

Arable', he was advised, when he began to farm in 1809, that 'the four course of Cropping was the best adapted to my farm'.[74] His arable land was comparatively good, and his permanent pasture was equal in area to about one-quarter of his arable[75] and his remark is correspondingly significant.

When Blaikie viewed the estate in 1816, he found that about three-fifths of the farmers whose rotations he noted (34 out of 56) had adopted the regular use of the four-course on the whole of their farms. Several of his comments pointed to a recent extension of the use of this rotation. He reported that at Billingford, some time between 1812 and 1816, Coke had given permission to the tenant to change from the six-course called for in his lease to the four-course, and two others in Billingford whose leases of 1803 called for a six-course had done the same. At Castleacre, Mr. Purdy, cultivating upon a six-course, 'wishes to adopt a 4 course'. At Dunton, a lease of 1811 prescribed a six-course but Blaikie found the tenant using the four-course. At Godwick, Blaikie considered that a tenant might be permitted, when his present lease expired, to change to the four-course 'with perfect safety, by reason of the large proportion of permanent pasture land'. Usually, however, Blaikie felt his task to be to restrain the spread of the four-course except on relatively good land. At Quarles, where the tenant was applying the four-course, 'the greater part of this farm is too weak soil to support a 4 course shift'. On two farms at Wighton the farmers had 'changed injudiciously of late years to the four course', and a third farm where the same change had taken place was described as 'over cropped'. At Weasenham 'a fine farm of good land has the appearance of being overworked by a four course shift'. Another farm in Weasenham evoked a general reflection from Blaikie:

> The four course shift is too frequently carried *to an excess.* To support such close cropping the land must either be of first rate quality or it must be supplied by extraordinary efforts in the collection and application of Manures. Where there is a large proportion of permanent Grass, or Cocksfoot layer [Blaikie was a great advocate of cocksfoot] for the support of Stock, a four course may be supported upon the Arable part of such farms. But when . . . a four course shift is attempted to be supported upon a large Arable farm (and although good) certainly not of the first quality and wanting permanent Grass, or cocksfoot layer, a sufficient quantity of manure cannot be produced.[76]

[74] 1816, L.B., 258.   [75] 1829, L.B., 31.   [76] Blaikie's Reports.

At the sheep-shearing of July 1821, Coke recommended the four-course on good land and the five-course on light land.[77] As late as 1831 Blaikie still regarded the four-course as by no means certainly desirable. He wrote then that 'the four and six course shifts taken alternately, is preferable to a constant repetition of four-course husbandry and should be adopted at every convenient opportunity'.[78] By 1851, on the other hand, the four-course was standard and accepted practice. Keary, in the report he produced in 1851, containing a description of every farm on the estate, wrote constantly of the 'usual four-course shift',[79] though he stated, even then, that the very lightest lands on the estate were kept under a five-course.[80]

Near the end of his life, in 1841, Lord Leicester still advocated the four-course and expressed rigid opposition to successive corn crops. His letter is worth quoting in full as a statement of the long-established husbandry doctrine of the elder statesman of English farming. By 1841 Lord Leicester's eyes were too weak to allow him to write, and his young wife wrote for him to Mr. Gibbs of Swansea, who had explained that he was helping to form an agricultural society and had asked for advice.

In reply to your Letter, Lord Leicester begs to say that it would be difficult for him to lay down any rules for Farming unacquainted as he is with your Soil, however he thinks he cannot err in recommending to the Notice of your Society Mr. Blaikie (his late Steward's) Works upon Agriculture, which are simple and instructive. Under any circumstances he advises your getting rid of fallows and substituting Turnips and Mangel Wurzel. If the Soil is genial he would recommend the four course husbandry, after Turnips, Barley laid down with Clover, after this the land to be sown with Wheat after one Ploughing in the Autumn, and upon Light Soils let the clover remain two years, manuring upon the surface and then plough for Wheat, and how ever fertile the Soil, never sow two White Crops in succession. He recommends the Drill Husbandry.[81]

Lord Leicester's restatement of the rigid Holkham prohibition of successive straw crops was already coming to be outmoded.

The standard non-animal manure in Norfolk for most of the first half of the nineteenth century was powdered oil-cake, from rape or

---

[77] *Norwich Mercury*, 7 July 1821.   [78] A.L.B., i. 154.
[79] e.g. Overman's farm, Burnham Sutton, Keary, i. 83-6.
[80] e.g. ibid., and Keary, i. 237-41, on Massingham.   [81] A.L.B., ii. 39-40.

linseed cake. In 1816, referring to Norfolk, Blaikie wrote, 'Oil cake is very generally sown with the Wheat Crop—it is a Cheap manure at the present prices, and wonderfully increases the quantity as well as improves the sample'.[82] In the same year General Fitzroy conducted experiments on fertilizers using oil-cake, lime, and malt coombs.[83] In 1831 Blaikie explained the use of oil-cake to Sir John Sinclair. In the alternate four- and six-course he advocated, both turnip breaks would be manured and one of the two wheat breaks in each complete cycle: three dressings of manure in ten years, of which powdered rape-cake alone would be used for the wheat, in addition to the effect of folded or grazed sheep. Other manuring rhythms he suggested for different rotations involved three or four manurings in nine years, or two manurings in five years. Blaikie included a highly significant remark on the manuring he advocated for the four- and six-course he demanded for the best light land:

> To a person unacquainted with the management of light arable land, and the use of Rape cake, it will appear the three dressings of manure mentioned here exclusive of the sheep fold is extraordinary high farming. But when the expense and speedy application of the manure is pointed out, the wonder ceases. Thus, the average cost price of Rape Cake including the expense of breaking the same into a powdered state has in the last ten years been about £5.10.0 a Ton and that quantity is usually allowed to three acres of land.

This rhythm of manuring would certainly not have been thought 'extraordinary high farming' in west Norfolk in 1851. By then, the practice of manuring for *every crop* was gaining ground. The cause was that it became easier and cheaper to buy fertilizers; the consequence was that it became possible to break away from the four-course rotation.

In 1850 Caird described Hudson of Castleacre, a Coke tenant, manuring for every crop. Overman of Burnham Sutton, another tenant, followed 'the same system of high cultivation', and Mr. Blyth, at Sussex farm, in the Burnhams, 'manures for every crop where he thinks it is required'. The artificial manures Caird found in use in north-western Norfolk were salt, nitrate of soda, and superphosphate, in addition to guano; he wrote of the direct application

---
[82] 1816, L.B., 128. Oil cake had been used in Norfolk as fertilizer even before Coke of Norfolk's time. See A. Young, *Farmer's Tour through the East of England* (1771), iv. 433.
[83] 1816, L.B., 126.

of artificial manures to every crop as having become a 'matter of system' since the year 1846.[84] On the other hand, Baruch Almack, in a careful account of Hudson's methods of cultivation at Castleacre, which he included in a R.A.S.E. prize essay probably written in 1842 or 1843, does not suggest that he had then taken to manuring for every crop.[85] It seems that Coke's death, in 1842, took place when the Norfolk four-course rotation had reached perfection, and when the causes of its modification were appearing.

New artificial fertilizers had been tentatively tried during nearly the whole of the first half of the nineteenth century, but their use was restricted. As early as the sheep-shearing of 1819, nitrate of potash was said to be 'beneficial to wheat'.[86] In 1818 Coke advocated the use of gypsum (calcium sulphate),[87] though for the Park farm at Holkham only oil-cake, town manure, lime, and soot were bought in 1817. In his closing speech at the last sheep-shearing in 1821, Coke said oil-cake was 'pre-eminent' as a manure, but referred also to gypsum, yeast manure, and bone manure whose 'use is extending in the county'.[88] It was not until 1839 that serious reference to the value of nitrates appeared. Then a Mr. Macdonald, apparently replying to a letter from Lord Leicester, wrote to say that his attention had been drawn some time ago to the value of saltpetre (potassium nitrate) as a manure.[89] In the same year, Lord Leicester asked a tenant about the effect of nitrate on his crops. The tenant (Garwood of West Lexham) told him that he had applied 8 stone per acre of cubic nitre (sodium nitrate) in the month of May to some wheat growing on poor land. The yield without nitre on the same land was 10 bushels less than from the portion treated with nitre.[90] Guano began to be imported during the last years of Lord Leicester's life, after about 1840.[91]

Evidently the aged Lord Leicester took an interest in the practical applications of new fertilizers, but he had little sympathy for

[84] A.L.B., i, 153–5. J. Caird, *English Agriculture in 1850–1*, 168–72.
[85] B. Almack, 'On the Agriculture of Norfolk', *J.R.A.S.E.*, v (1845), 322–30.
[86] Speech of Mr. Holditch, *Norwich Mercury*, 10 July 1819.
[87] Ibid., 13 July 1818.
[88] Genl. Receipts and Payments, Miles Bulling, Holkham farm. (Game Larder.) R. N. Bacon, *A Report of the Transactions at the Holkham sheep shearing* . . . (1821). In 1819 a manufacturer of salt asked for a testimonial from Coke of the value of salt as a manure for wheat and potatoes. Blaikie replied that Holkham got enough salt from sea breezes. A.L.B., i. 4–7.
[89] A.L.B., ii. 30–2. [90] Ibid., ii. 33.
[91] G. R. Porter, *Progress of the Nation*, 3rd edn. (London, 1851), 144n.

The estate under Coke, 1816-1842

attempts to work out their theoretical basis. Towards the end of his life, his daughter wrote, at his dictation, a sad letter to Lord Western. The old hero of English agriculture was depressed by its prospects:

> ... my Father thinks that Agriculture stands quite as bad a chance and that the Society with the Duke of Richmond and Lord Spencer at its head[92] will do a great deal more harm than good, as it does not lead to improvement in any way, nothing new is elicited and he considers all the attempts to introduce Chemistry as an Engine in Cultivation as a complete fallacy, Sir Humphrey Davy having had full power to exercise his skill in that Science for so many Years here without producing one Shillings worth of benefit. My Father's friend Mr. Handley was much taken with the idea and had a Chemist staying with him for 3 weeks without any satisfactory result and all he could say for himself was that Mr. Handley's Soil did not suit his Chemistry, a fact Mr. Handley did not attempt to deny, when my Father told him the story here this Winter —It is upon these grounds that my Father has declined belonging to any of the Agricultural Societies in England, more especially to the General one, the Meetings of which seem only concerned to make speeches and compliment each other, but he is an Honorary Member of all the Societies in Scotland which he thinks are productive of greater benefit and that in spite of the disadvantages of Climate that Country will soon have the Agriculture of England quite in the background...[93]

v. *Tenants' cropping and the size of farms*

The progress made in the farming on his estates in the last fifty years or so of Coke of Norfolk's time can perhaps be illustrated best by comparing what was being grown on two farms early in the 1790s, with their cropping in the early 1850s. Unfortunately no evidence to show what tenants were doing exists for any years nearer to the time of Coke's death in 1842, and it is very likely that the years 1842-51 saw some further advance towards 'high farming'. The tables below must not be thought of as indicating precisely the state of affairs of 1842, but we can safely assume that most of the changes they show came before 1842.

First, a farm at Holkham, the Branthill farm. In 1789-91 it had 918 acres arable land; in 1851-3 it had 905. Table O shows the acreages of the various crops.

[92] Lord Leicester meant the Royal Agricultural Society of England.
[93] A.L.B., ii. 37, 10 Feb. 1841.

TABLE O

	1789			1790			1791		
	a.	r.	p.	a.	r.	p.	a.	r.	p.
Wheat	64	0	0	146	0	0	130	0	0
Barley	206	0	0	151	0	0	181	0	0
Oats	48	0	0	31	0	0	46	0	0
CEREALS	318	0	0	328	0	0	357	0	0
Ley	358	0	0	277	0	0	244	0	0
Turnips	171	0	0	181	0	0	158	0	0
Fallow	71	0	0	132	0	0	159	0	0
TOTAL	918	0	0	918	0	0	918	0	0

	1851			1852			1853		
Wheat	243	2	26	187	3	37	233	2	34
Barley	154	1	3	247	1	5	184	1	14
Oats	67	1	32	25	2	27	14	3	16
CEREALS	465	1	21	460	3	29	432	3	24
Ley	192	3	6	187	3	38	259	0	15
Roots	247	1	36	209	3	6	213	2	24
Fallow	—			46	3	30	—		
TOTAL	905	2	23	905	2	23	905	2	23

A great extension of the acreage under cereal crops is evident—in particular, there was much more wheat in 1851. The length of leys has been shortened. Fallows have almost disappeared, to be replaced by an extension of the area carrying roots. The four-course has taken over. There was very little pasture or meadow on this farm and the soil was extremely varied, some of it quite good, much of it very light and sandy.[94]

The second example is from Castleacre: the famous Wicken farm of which it was said in 1851 'There are perhaps few farms which possess so many desirable properties as this, fine land, large well-laid out fields.' Some of the soil was thin and poor, some of

[94] Keary, i. 5–6.

The estate under Coke, 1816-1842

it rather stiff loam on clay, some of it 'a deep good mixed soil upon excellent marl or clay, very fertile and productive'.[95]

In 1789 there were 974 acres of arable with 162 of meadow and pasture; in 1851 there were 1058 a. 2 r. 16 p. of arable and 151 a. 0 r. 27 p. of meadow and pasture. The cropping is shown in Table P.

TABLE P

	1789			1790			1791		
	a.	r.	p.	a.	r.	p.	a.	r.	p.
Wheat	150	0	0	106	2	0	222	0	0
Barley	308	2	0	258	0	0	197	0	0
CEREALS	458	2	0	364	2	0	419	0	0
Ley	287	2	0	229	0	0	301	0	0
Turnips	156	0	0	118	2	0	154	0	0
Peas	32	0	0	154	0	0	—		
Vetches	26	0	0	—			—		
Fallows	14	0	0	108	0	0	100	0	0
TOTAL	974	0	0	974	0	0	974	0	0

	1851			1852			1853		
Wheat	400	0	5	293	3	23	226	0	8
Barley	140	3	13	223	3	15	314	3	10
CEREALS	540	3	18	516	6	38	540	3	18
Ley	255	3	23	226	0	8	261	3	15
Roots	261	3	15	314	3	10	255	3	23
TOTAL	1,058	2	16	1,058	2	16	1,058	2	16

Once again, the four-course has taken over. This time the growth in the area devoted to cereals is less marked than on the more consistently light soil of Branthill. The greater acreage devoted in 1791 to wheat compared with barley was unusual at that time—it was probably a result of the high quality of some of the soil of the Wicken farm. In this instance, fallows have completely disappeared—

[95] Keary, i. 223-4, Dr. J. E. G. Mosby, *Norfolk*, part 70 of *The Land of Britain, The Report of the Land Utilisation Survey of Britain*, ed. L. D. Stamp (London, 1938), 217-20, describes the Wicken farm and gives its cropping for 1922-37 and in part, for 1851-5.

replaced by a larger acreage devoted to roots. The cropping of both farms demonstrates the late date of the complete triumph of roots over fallows.[96]

The death of Coke coincided with the coming to maturity of the four-course, an elegantly self-contained rotation. Keary's report of 1851 was a tribute to the good condition of Coke farms, nine years after Coke of Norfolk's death. There was only one substantial exception: the Egmere farm, near Holkham, of 1,122 acres, was a fine occupation with 'immense capabilities ... almost entirely thrown away'. There were few swedes and no mangolds; the swedes and the remaining roots were overrun with twitch and every other sort of weed. The wheat stubbles were 'as foul as possible'. The grass was under-grazed for there was hardly any livestock. The fences were dilapidated. Keary detected an 'entire absence of capital and an utter want of energy and skill'. The ferocity of his strictures on the farmer at Egmere are a measure of the standard attained by the other farmers on the estate.[97] In 1851 the whole estate contained about 43,000 acres. Table Q shows the comparative sizes of the farms on the estate and those of 1780:

TABLE Q

Size of farms in acres

	5-49	50-99	100-299	300-499	500+
1780	25	5	23	18	18
1851	25	6	19	17	34

The number of relatively small farms was little reduced in Coke of Norfolk's time. The substantial increase in large farms which took place—in 1851 there were six farms larger than 1,000 acres compared with four in 1780—seems to have been more the result of purchases and the acquisition of newly enclosed common lands rather than of consolidation of smaller farms.[98] Large farms characterized the estates before Coke came to Holkham. The principal argument Blaikie put in 1831 for large farms had applied throughout the previous century: 'That is The Sheep Fold ... That admir-

[96] Field book, 1789-1802; field books, Holkham and Castleacre (Estate Office).
[97] Keary, i. 97-9.
[98] G.E.D., 77, Keary, i and ii. The figures for 1780 are defective in that in two divisions, including about 2,500 acres of arable land, the holdings of tenants are not listed.

able husbandry is indispensably necessary in maintaining light Arable land in a good state of cultivation more particularly when situated at a distance from Manure and Market, and a flock of sheep cannot possibly be kept to advantage upon a small farm.'[99]

vi. *The condition of labourers*

The condition of farm-labourers did not occupy much of the attention of the Cokes and their advisers until the discontent of the labourers obtruded itself in the 1830s. Even the steady rise in the poor rate over some decades aroused less alarm than the manifestations of the 1830s, though rates were a formidable burden on tenant farmers. Blaikie asked certain farmers in 1816 what proportion the rates they paid bore to their rent. Their replies varied between 2*s*. 11*d*. and 7*s*. 3*d*. per pound of rent paid, for the years 1813–16.[100] Such sums reduced landlords' income by making tenants less able to pay high rents. The situation grew worse after the war: in 1833 a tenant declared that the 'progressively increasing Poor Rate' had 'more than doubled since I first came here (now 20 years)'.[101] Before 1833 the sufferings of the labourers had found an outlet in violent demonstrations.

Towards the end of 1831 the labourers of Burnham Overy and Burnham Thorpe, villages only two miles or so from Holkham, gathered together and equipped themselves with sledge-hammers. With these, they began their protest by breaking a threshing machine in Burnham Thorpe and then went on to smash the threshing machine belonging to Coke's tenant in Burnham Overy, before setting off to another farm to extend their depredations. By this time word of the insurrection had reached Holkham. Coke's already advanced age did not prevent his taking decisive action; his courage and determination are almost incredible in a man no less than 77 years old. No sooner had the news arrived, than Coke rushed to horse and set out, leading a cavalcade consisting of Blaikie, Blaikie's assistant, the estate office clerk, the farm manager, and the estate 'architect'. They were reinforced on the way by a few others, mostly tenants, until they were fifteen or so in numbers, before they came upon 100 to 150 rioters armed with large bludgeons in addition to their sledge-hammers. In Blaikie's words, 'They had a formidable appearance. They were halted with some

[99] A.L.B., i. 171-2.   [100] 1816, L.B., 101, 123, 133, 138.
[101] 1833, L.B., 134, from Hastings of Longham.

difficulty, Mr. Coke addressed them with kindness but without effect. The King's proclamation for the suppression of such assemblages was then read, but no better effect. The misguided men persisted in proceeding on their work of destruction and threatened the lives of all those who should venture to oppose them.' Coke promptly charged the mob and scattered them. The ringleaders were picked out and Coke set the example in attack by seizing the first man by the collar. This man was the principal leader, but he was apparently soon released because of his previous good character. Four others, however, were seized, after a short scuffle and an attempt at rescue, and these four were bundled into Coke's carriage, which had been brought along in the rear of the forces of order; three of the rioters sat in the carriage with one constable and the fourth shared the dicky with a second constable. Blaikie commanded an escort of four horsemen and, under their guard, the prisoners were safely delivered to the Bridewell in Walsingham. Coke and his other horsemen dispersed the remaining rioters before they left the scene of action. Blaikie was able to report to Westminster that Coke's action had had 'a most beneficial effect in the neighbourhood' and that 'no tumultuous assemblages of farm labourers have since taken place'.[102]

Rick-burning and other forms of arson broke out around Holkham. In December 1831 Coke bought two bloodhounds to use in tracking incendiaries and in January 1832 he subscribed £20 to the Freebridge and Lynn Association for preventing Arson.[103]

The Norfolk witness before the select committee of 1833, part of whose evidence was quoted earlier, gave details of the social condition of the county at this time. Here are three questions put to him and the answers he gave:

'Do you think that the uncomfortable situation of the tenantry, arising from that apprehension with respect to the conduct of the labourers has deterred peaceable and respectable men from going into farming?'—'No doubt of it; any man at all above the rank of a farmer would not incur the nuisance of it.'

'That struggle between keeping up profits by beating down wages is so painful, that men are indisposed to embark in it?'—'Yes, and the nuisance of having a quantity of unemployed people teasing them for allowances and threatening them.'

[102] 1831, L.B., 115-16, Blaikie to C. Bouchier, 28 Dec. 1831.
[103] 1831, L.B., 100, 106; 1832, L.B., 3.

'Is there the same good feeling now between the farmers and the labourers that there was formerly?'—'Nothing like what there was before the fires.'[104]

The implication was that men with large capital were less inclined to go into farming than previously, and that the size of farms was therefore falling. There is no sign, however, that it was difficult to find tenants for the farms of the Coke estate, or that there was any need to reduce the size of farms to attract prospective tenants. In 1831 the Holkham estate office still had a list of candidates for farms.[105]

The new Poor Law was welcomed at Holkham. Coke himself told a correspondent early in 1836 that 'the present Government, the only one that ever had courage to grapple with the Poor Laws, are entitled to the greatest praise from all Agricultural Proprietors, as well as the Poor, who will ultimately benefit by it'.[106] Blaikie's successor as agent, Baker, wrote later in 1836 that 'we find the New Poor Laws begin to work well in the districts where they have been established any time by making those men, absolutely employed, more subservient to their employers than heretofore, as they now find they have more to depend on their good conduct and Industry than has hitherto been the Case under the Old System'.[107] Another effect of the Poor Law was to set off an 'emigration mania' in northwest Norfolk. 'Now they are leaving this Neighbourhood to a very great extent, which I believe the operation of the New Poor Law Bill has been the principal cause of, the lower class of people having absolutely taken fright at it.'[108]

There was little industrial development in nineteenth-century Norfolk, and this made the problem of poverty more acute. Baker wrote to a correspondent in Manchester: 'The greatest drawback we have is the superabundance of Agricultural Labourers which press hard upon the occupiers of the Soil—I often wish we could transport a few into Your district, as we learn from report that you stand in great need of them, and are giving high wages.' Some did go off to look for work in other parts of England: 'We have lately sent from one Parish 26 Young Men into Northamptonshire to seek employment on the Rail Roads.'[109] Others set off overseas and for a time Coke contributed to their expenses—he paid one-third of the expenses in parishes where he was sole proprietor, and in proportion to his ownership where he was not. When he occupied

[104] Parl. Papers 1833 (H.C.612), V, p. 104, qs. 2220, 2222-3. [105] 1831, L.B., 28.
[106] 1836, L.B., 16. [107] 1836, L.B., 97-8. [108] Ibid., 38. [109] Ibid. 98.

some of the land he owned, he contributed a larger share. Apparently the parish incumbents organized the emigration parties, as at Mileham, where about twenty adults and twenty children were to leave for America in the Spring of 1836, and in Kettlestone, where thirty-five persons were preparing to embark for Canada.[110] But doubts of the wisdom of encouraging emigration soon appeared. Baker wrote in March 1836: 'As to the consequences of Emigration, I have my fears that it may be carried too far, particularly as you cannot choose your Emigrants, and it is beyond doubt that the best labourers and most Industrious persons are the people now leaving the Country', and towards the end of 1836, Coke decided not to contribute further help towards paying for emigration.

The unattractive conditions in which farm-labourers lived emerge vividly from Keary's report on the Coke estate in 1851. His description of the eight cottages let with the Harpley Dam farm in Flitcham is a fair sample of his remarks on cottages; the condition of farm cottages throughout the estate compared unfavourably with that of other farm buildings.[111]

Occupiers	Description	Remarks
1. Amy Sayer and Thos. Sage	Cottage and yard	2 families in one house, 5 persons. In bad repair.
2. John Sayer	do.	2 bedrooms, 9 persons. In bad repair.
3. Widow Fulcher and 2 children and Thos. Fulcher and wife	do.	2 bedrooms, 5 persons. In bad repair.
4. Robert Harrison	do.	1 bedroom, 3 persons. In bad repair.
5. Robert Candler	Cottage	1 small bedroom, 7 persons. Very small and bad indeed.
6. Robt. Large and wife and 1 grown-up son and Thos. Large, wife and 4 children	do.	2 very small sleeping rooms and unfit to live in.
7. William Harrison and wife and Robt. Harrison, wife and 2 children	Cottage and garden	2 very small bedrooms. In fair repair.
8. William Piggins and 7 children	do.	2 very small bedrooms. In fair repair.

It is unlikely that the farm-labourers shared the respect, admiration, and gratitude that the tenantry felt for their famous landlord.

[110] 1836, L.B., 36-8, 49. [111] Keary, i. 146.

11

The Park Farm—the Post-War Crisis and after

THE years after the great war were years of agricultural depression, culminating in the crisis of 1821-2. Prices for farm produce, which had been at a high level from 1805 to 1814, were rather lower in 1815, but recovered and became very high once more in 1817. Thereafter, prices fell steadily, reaching very low levels in 1822 and early 1823. In particular, prices immediately after the 1822 harvest were calamitously low.[1] Prices rose again in 1823 and 1824 and held up until the middle 1830s.

Prices for sheep were low in 1821 and 1822, as well as those for grain, though cattle prices remained reasonably good.[2] The total flock of the Holkham farm of 2,004 sheep were valued at £2,810 on 1 January 1817, while 1,786 sheep were valued at only £1,836 on 1 January 1823. In September 1819 George Tattersall paid Coke 28s. for each of 200 lambs; in July 1821 Thomas Leeds paid 20s. for each of 200 half-bred (presumably Leicester crossed on Southdown) lambs; in August 1822 he paid 14s. each for 200 half-bred lambs and only 11s. each for 160 more. Coke allowed him 3s. rebate on each of the 200 lambs he had bought the year before 'for loss in Bargain'.[3] No doubt farmers were failing to replenish their flocks, or throwing them on to the market to raise money to cover their commitments. At the sheep-shearing of July 1821 there were apparently no bids at all for sheep offered for sale.[4]

The effect of the slump on the Park farm is best shown by the following figures of profit and loss for the years 1817-26.[5]

---

[1] T. Tooke, *A History of Prices and of the State of the Circulation from 1793 to 1837*, 2 vols. (London, 1838), i. 390, table of wheat prices, 1793-1837.
[2] 'Stock book 1814' (Game Larder).
[3] Ibid.
[4] *Norwich Mercury*, 7 July 1821.
[5] 'Farm Accounts' (Game Larder).

		£	s.	d.			£	s.	d.
1817	Profit	548	2	10	1822	Loss	2,736	19	0
1818	Loss	797	13	0	1823	Profit	119	1	4
1819	Loss	3,081	1	5	1824	Profit	971	11	$8\frac{1}{2}$
1820	Loss	967	9	8	1825	Profit	486	18	8
1821	Loss	5,095	11	$2\frac{1}{4}$	1826	Loss	1,207	0	$1\frac{1}{2}$

These figures were worked out at the end of each calendar year, taking into account a valuation of stock and crops at the beginning and end of the year.[6] They are incontestably depressing; it is not surprising that Coke's daughter, Elizabeth, should have written to her betrothed from Holkham in November 1822, 'I hope you will have as little to do with farming as possible. Just enough for one's own wants and to supply the ménage—further than that it is a most expensive amusement.'[7] Indeed, this remark, and the figures above, suggest the possibility that the Park farm was an uneconomic, extravagant showpiece, whose technical beauties were secured at an expense which would have been impossible for a tenant depending on his farm for his livelihood.

The scale of operations at Holkham was enlarged in 1823, when the Quarles farm was taken into hand, probably adding 500 acres or so. Thereafter the farm remained the same size for the rest of Coke's lifetime, though it is possible that a larger part of it was devoted to plantations as time went on. In 1839 Lord Leicester, as Coke had now become, occupied 3,762 acres of the 5,208 acres of the parish of Holkham (all of which he owned, apart from 6 acres in other hands and 24 acres of public roads). Of the area in Lord Leicester's hands, 1,209 acres were arable, 541 acres were pasture and 1,155 acres were woodland.[8] In 1821 the farm had ceased to be publicized by its annual agricultural show—the Holkham sheep-shearing of July 1821 was the last. Probably the sheep-shearings were brought to an end as one of the measures of economy that appeared to be imperative in the early 1820s; they must have seemed especially extravagant functions when bids for sheep were no longer made at them. It seems that another method of spread-

---

[6] The 1820 loss was comparatively low only because Blaikie had been supplying funds to the farm manager from time to time of a total of £2,400 'to carry on Farm'.

[7] A. M. W. Stirling, *The Letter-Bag of Lady Elizabeth Spencer-Stanhope*, 2 vols. (London, 1913), ii. 54.

[8] Holkham Tithe Account, 1839 (copy in Estate Office). There may have been some additional land in hand in Wells parish; but if so, it would include very little arable.

ing knowledge of Holkham technique came into existence in this period. The Park farm manager lived in Longlands farmhouse, at Holkham. The local enumerator for the 1841 census recorded the residence there of four 'agricultural pupils'. They were all young men of 20 from other counties.[9]

The time of striking innovation, however, was past; as, probably, elsewhere in the area, growing intensification of cultivation, higher farming, and increasing care in the management of stock and crops were no doubt the main features of the Park farm after the war. The high farming of the 1850s and 1860s rested on cheap and abundant labour and on new fertilizers. The new artificial fertilizers and guano were only beginning to be used by the 1840s;[10] but cheap labour, on the other hand, was embarassingly plentiful in north-western Norfolk after the war. The Park farm possessed two threshing machines (valued at £50 each) at the beginning of 1817, but by 1822, Blaikie was writing that 'Thrashing machines and all other Agricultural implements calculated to abridge manual labour are now very generally falling into disuse'.[11] The folding of sheep on unfenced parkland could hardly have been extensively practised without cheap labour: in 1851 Keary reported that as 'many as 500 dozen Hurdles' were sometimes simultaneously in use on the Park farm.[12]

It is unlikely that the four-course rotation was used on the Park farm as early as 1816. Blaikie's cautious suspicions of its desirability, which he still expressed as late as 1831, have been recorded earlier.[13] In 1850, by contrast, Caird wrote of the Park farm being 'managed in the usual four-course rotation', but by that time guano and superphosphate were in use.[14] In 1817 the only fertilizing agents bought from outside were oil-cake, soot, and manure. In that year 521 loads of manure were bought for the farm from Wells, 506 bushels of soot, and just over 100 tons of oil cake. Forty-four tons of gypsum were bought in 1819, but the position in 1826 is very much like that of 1817: 453 loads of muck were bought from Wells and 118 tons of rape cake and $57\frac{1}{2}$ tons of linseed cake were bought.[15]

[9] P.R.O., HO 107/772.
[10] R. N. Bacon, *Report on the Agriculture of Norfolk*, 111.
[11] 'Farm Account, 1817-26' (Game Larder); A.L.B., i. 71.
[12] Keary, i. 12-13.   [13] 1816, L.B., 128; A.L.B., i. 154.
[14] J. Caird, *English Agriculture in 1850-1*, 167.
[15] 'Farm Accounts, 1817-26' (Game Larder).

In 1826 the Park farm produced about 109 lasts of barley and about 108 of wheat.[16] If various assumptions are made the yields of 1826 can be worked out to have been about 9 coombs per acre of both wheat and barley—which would compare very well with the yields of the 1780s. But the figure is not very reliable.[17] That the wheat estimate may be rather high is suggested by a letter from Coke of 5 September 1832. He wrote to Lord Lynedoch, 'You will be surprised to hear after the rains . . . that I should have . . . the most abundant, and the most superior quality of corn I ever grew in the finest possible condition. I think I do not hazard too much when I say that my wheat crop which consists of 330 acres will not yield less than 40 Winchester bushels per acre' (i.e. 10 coombs at 4 bushels to the coomb).[18] The roots acreage in 1826 seems to have been about 245 acres (beet, swedes, red turnips, and white turnips).[19]

At the end of 1826 the livestock on the farm was valued at over £7,800. It comprised 78 horses and colts, 277 head of cattle, 2,751 sheep, and 179 pigs. At the beginning of 1817 farm implements were valued at about £1,000. They included two threshing machines, four winnowing machines, seven drills, two machines for sowing broadcast, and fourteen wheel ploughs. Seeds in store included cocksfoot, trefoil, white and broad clover, green round turnip seed, and Swedish turnip seed. The total value of livestock, crops, and implements then was put at over £13,000—a figure which shows how large a capital was needed for working a big Norfolk farm.

Unhappily no account book of the home farm survives for a time later than the 1820s, so that it is not possible to discover what was going on there in the last decade of Coke's long life. There are some

[16] A last is 20 coombs or 80 bushels. The production figures are approximate. They are based on changes in stocks held between January 1826 and January 1827 and on the quantity sold in 1826. Stocks of wheat and barley are recorded as 80 lasts each on 1 Jan. 1826—alarmingly round figures. The caution again applies that grain used for the activities of the farm department of the Coke administration might not be recorded—though seed corn, at least, was being bought for the farm by that time. The harvest of 1826 was above average for wheat; but there was drought, which usually does damage in north-western Norfolk (T. Tooke, op. cit., ii. 137).

[17] It depends on three assumptions: that the total arable acreage was the same as in 1839, that the acreage of wheat was the same as that of barley, and that the proportion of arable land under wheat and barley was precisely two-fifths. If a five-course rotation (with a two-year ley) were being followed the last assumption would be reasonable, but if a four-course were being used the figures of yields above would be too large.

[18] National Library of Scotland, MS. 3620, f. 357. Coke's writing is difficult to read, and his claim for his barley cannot be reproduced with complete confidence—it appears to be 15 coombs (60 bushels).

[19] This figure fits the assumption of a five-course rotation quite well.

signs, however, that the rearing of stock reached steadily higher levels of achievement. In December 1834 a London butcher wrote to Coke to congratulate him on a 'most extraordinary Devon bullock . . . I have seen and killed very many most beautiful Bullocks but not one so good in every respect as the one I am now in possession of, and allow me to say (Sir) it must be very much to your satisfaction to know you have both *Bred* and *Fed* the best Bullock ever seen'. The writer slightly spoiled the effect by a postscript: 'I trust I shall not be Disappointed having your orders.'[20] A year later, another butcher wrote with similar congratulations on two bullocks, from which he sent Coke pieces of meat.[21] Coke continued to devote attention to sheep: in 1837 he wrote in his emphatic way about his latest improvement in sheep breeding to the Revd. Mr. Bennet,

> You may recollect my telling you at Swaffham (altho I had never seen him) that if your Sheep was superior to mine he must be a very good one. And that the Cross into a Hampshire ram would beat the *pure Down*. I have had sufficient experience to convince me of this, and unless Breeders, both of Cattle and Sheep, will divest themselves of prejudice, they must always remain in the background—Prejudice is the Bane to all Improvement.[22]

In his later years, Coke improved the Suffolk breed of pigs by crossing them with Neapolitan pigs.[23] In the last months of Lord Leicester's life, a request went to the Treasury from Holkham for permission to import four pigs then being shipped from Naples 'for the improvement of the breed in England'.[24]

The state of affairs on the Park farm at the end of Lord Leicester's life is illustrated by the inventory of its stock and crops taken after his death in the summer of 1842.

There were then 283 acres under wheat and 270 under barley together with 40 acres of peas and 4 of vetches, 234 acres of beet and turnips and 270 acres of new layers. There were 12 stacks of hay and fodder. There were $1\frac{1}{2}$ tons of linseed cakes in store, and 5 cwt. of saltpetre. The livestock on the farm consisted of 2,181 sheep and lambs, 150 head of neat cattle, nearly all Devons, 40 horses and colts, and 98 pigs.

---

[20] A.L.B., i. 210.   [21] Ibid., i. 223.   [22] Ibid., i. 221.
[23] Earl Spencer, 'On the Improvements which have taken place in West Norfolk', *J.R.A.S.E.*, iii (1842), 5.
[24] 1842, L.B., 41.

The figures for crops show that the four-course rotation was in use. We can be sure that the Park farm, strengthened by its unusually high proportion of pasture land, and by the generous use of labour, and fertilizers brought in from outside, was a thing of great beauty. It is unlikely that a farm of such prestige was much affected by the financial crisis which is examined in the next chapter of this book.

12

General Finance, 1822-1842

i. *The Lancashire estate and Coke's coal-mine*

IN this Chapter the effect of the post-war crisis on Coke's finances will be analysed. As a preliminary the curious story of Blaikie's attempt to establish Coke as a coal-owner and the closing stages of Coke's career as a lighthouse-keeper will be described.

Before Blaikie came to Holkham, Coke evidently failed entirely to grasp the possible significance for his future income of industrial development: the benefits he or his children might derive from land situated in growing towns or on coal measures. It was Coke of Norfolk's actions that determined that his estate should be purely agricultural, a Norfolk estate, cut off, after the lighthouse was taken away, from the profits coming from manufacture, trading, and mining, and the towns they created, profits which more far-seeing landowners, or landowners whose property was more fortunately situated, often shared in, and which kept up their incomes when the triumph of British industry was followed by disaster to British arable farming.

Coke's London properties were sold in 1786 in order to help to pay for lands bought in Norfolk. The Lancashire estates were sold between 1790 and 1804, partly in order to pay for still more Norfolk land; all the estates sold were near Manchester. For some reason one small estate survived: a property in Reddish let to one Hyde by Wenman Coke in 1771 on lease for ninety-nine years for £125 a year rent, and 'a valuable consideration' at the granting of the lease. This lease was illegal under the settlement of the Lancashire estates, since it was not made determinable on three lives.[1]

Blaikie's attention was first drawn to the Lancashire estate, about which he had 'previously heard but little', by a letter which came to Coke from Manchester in 1823. It came from a Mr. Ashworth, who was acting for some persons who wished to purchase the mineral on an estate of 40 acres in Pendlebury, near Manchester,

[1] 1823, L.B., 117; 1824, L.B., 163-4.

one of the estates Coke had sold. It seemed that the mineral had not been sold with the land. Blaikie made careful search in the estate office, but could not discover any documents relating to the Lancashire estates. Hanrott, in London, knew nothing about them. As to Coke himself, Blaikie wrote crossly, 'On all these important matters Mr. Coke appears either to have had no particular information, or will not *now* give himself the trouble of casting the subject over in his mind, and making search for such documents relating thereto as remain in his possession.' Blaikie did not share his master's indifference for he understood that if Coke still owned part of the Lancashire coalfield, the profits might be great. Blaikie wrote to Ashworth and to Newton, the Derby solicitor who had acted for the Cokes in their dealings in Derbyshire and Lancashire, to ask them to find out what they could. Ashworth reported that Coke had once held a 'very considerable' property in the neighbourhood of Manchester: 'Tetters Fold, the township of Reddish, some property near Chetham Hill and Pendlebury'.[2] Newton sent depressing news. He had found several draft conveyances, and all of them expressly included minerals or contained words comprehensive enough to include them. There was only one exception—the conveyance of about 20 acres of land in Pendlebury had reserved the minerals to Coke.[3] In the spring of 1824 Ashworth wrote to ask the price for which Coke would let those minerals.[4]

Blaikie decided to go to Manchester himself to investigate the situation and he travelled there in September 1824. Mr. Ashworth and his five sons impressed Blaikie deeply: 'I have seldom met with a more shrewd sensible and industrious family.' Ashworth was engaged in cotton-spinning as well as in land-agency, and he had acquired a 'very beautiful' estate adjoining his mills. Three of his sons were brought up to 'trade' and two to 'agency' and 'neither he nor his sons ever stir a step without being paid 3Gs. [Guineas] a day each exclusive of all expenses.'[5]

Blaikie and the Ashworths found estate after estate sold by Coke, often by carelessly drafted conveyances, about one of which Blaikie wrote to Coke with incredulous horror 'there can be no doubt of your signature . . . being genuine'. Many of these estates were now known to be 'full of coal'. Sadly, Blaikie wrote to Coke, before setting off to Middleton, near Stockport, to 'see an estate which

[2] 1823, L.B., 106.   [3] Ibid., 228.   [4] 1824, L.B., 71.
[5] Ibid., 131.

you sold' where he hoped coal might have been reserved at the sale: 'I wish I had been here 20 years ago, I am even now not without hopes of picking up a little remnant for the young Lord Chief Justice', Coke's infant son and heir.[6]

Blaikie could salvage little. An attempt to persuade a waterworks company to pay for the minerals that might exist under their reservoir was not very rewarding. The tenant Hyde was found to have neglected his land to an extent that shocked Blaikie. He had sublet to two farmers paying, together, £250 a year more than the rent Hyde paid to Coke. The words of Blaikie's condemnation epitomized Holkham doctrines on the proper relation between landlord and tenant: 'Had Hyde enabled his tenants by means of moderate rents, and due encouragement on his part, to have improved the Estate by underdraining and other good husbandry practices, he might then have applied to you with a better grace for permission to transfer his interest in the lease. But he has omitted to do what he ought to have done in those respects—and he has done what he ought not to have done, that is, he has squeezed exorbitant rents out of the Tenants, so as to incapacitate them from effecting necessary improvements. So that your Estate, instead of being improved as it ought to have been in the fifty-six years it has been in Hyde's possession, has every appearance of having been greatly deteriorated.'[7] But Hyde's lease could not easily be overridden, however questionable its origins. Higher hopes were entertained of another transaction. Having confirmed that the mineral had been reserved when two small estates in Pendlebury were sold, Blaikie agreed to let the coal to 'a large coal proprietor in the neighbourhood', Andrew Knowles. Knowles agreed to lease the coal for twenty-one years if it should last so long. He was to pay a minimum of £300 a year while the coal lasted, except in the first year, when he would be sinking his pits, but the maximum rent was to be calculated on the quantity of coal extracted each year. Blaikie hoped that Coke in the course of the lease would receive about £4,500 for the coal Knowles would work from one seam 4 feet thick. There would remain, deeper down, a seam of 2 feet 8 inches which might become worth working later on.

Towards the end of 1825, Ashworth reported to Blaikie, 'A. Knowles, I informed thee before was sinking a Coal Pit which he has finished, and now driving the levels and getting some Coal,

[6] Ibid., 128-9.   [7] Ibid., 129-31.

in the course of a few weeks I expect he will be getting a considerable quantity.'[8] In fact, the results were disappointing. In the period ending in September 1836, Knowles paid to Coke £3,363. 9s. 11d. (after deduction of Ashworth's fees). Only rarely did Knowles win enough coal to bring the rent above the minimum £300 a year.[9]

Coke had cast away his substantial property in Manchester in a remarkably casual way. He and his advisers had clearly had no idea how valuable it might become, or even that it would be worth while concerning themselves with the minerals his estates might contain. This episode shows how much the growth or decline of the prosperity of a landed family could depend on accident. If Coke had not felt it necessary when he did to sell the Lancashire estates, his wealth would have grown to a much greater degree. If he had merely insisted on reserving minerals in the sales, large profits might have been gained. The fact that the central block of Coke's properties were in the purely agricultural county of Norfolk meant that his properties outside Norfolk were the first to go if sales imposed themselves. Coke therefore lost a chance of preserving for himself and his descendants a share in the rising value that industrial and commercial development brought to suitably placed estates. Other landowners, whose main properties were in industrial areas, no doubt devoted themselves to their extension and made any sales that were necessary from their detached properties, which were often in purely agricultural areas. In the nineteenth century wealth often came to landowners as the gift of geographical good fortune rather than as the reward of hard work or efficient management.

ii. *The loss of the Dungeness lighthouse tolls*

During Coke of Norfolk's time the value of the lighthouse at Dungeness—an essential element in his finances—steadily increased. From the 'yearly rents'—the lighthouse dues received from shipping less agents' commission—there were deducted the 'outrents', the small annual rent of £6. 13s. 4d. paid to the Crown plus an even smaller charge for rates; taxes—the Land Tax and, when applied, the Property Tax; the 'Bailiff's Fees'—the salaries for the attendants at the light and for the expenses of the head agent in London; and 'Repairs'—the cost of keeping a light burning; what remained was

---

[8] 1824, L.B., 129; 1825, L.B., 162.   [9] 1839, L.B., 22.

Coke's. Here are the figures, to the nearest pound, of the average receipts each year over periods of ten years in Coke's time:

	Yearly 'rents'	Net money
1778–87	3,401	3,108
1788–97	4,461	3,743
1798–1807	5,875	5,055
1808–17	6,841	5,738
1818–27	7,440	6,567

After the negotiation of a new lease the light tolls ceased to pay six or seven thousand pounds a year, and, in the years 1829–35, the average annual net return from the tolls was only £2,268.

The grave disquiet caused at Holkham by the prospect of the loss of the lighthouse is readily understandable. Until such threats arose, the management of the lighthouse involved little effort or attention at Holkham; Trinity House laid down its requirements and Coke's agents complied with them. No knowledge of lighthouse technicalities was needed from Coke or his servants at Holkham. In 1790 Trinity House demanded that a completely new lighthouse, of a type they specified, should be constructed at a point 100 yards from the sea at low-water—the old lighthouse stood 540 yards further north and an accretion of shingle had caused an extension to Dungeness Point so that the old light, especially in foggy weather, had lost much of its usefulness.[10] Although the existing lighthouse was substantial, as Coke later pointed out, the request was complied with.[11] A lighthouse was built in 1791 by Samuel Wyatt, at a total cost of £3,059. 0s. 4d.[12] Thus a completely new lighthouse was paid for out of one year's net revenue from the tolls charged to shipping.

Otherwise, there is little information about the lighthouse until after the Napoleonic war.[13] In 1817 trouble arose, but it was dealt with without much effort being necessary at Holkham. In September 1817 Commander Matthew Popplewell, R.N., was returning from Jamaica as a passenger in the *Fortitude*. He stayed up late on the fine balmy night of 24 September and gazed at the

---

[10] Mr. Dowsett, Principal Keeper of the Dungeness lighthouse, was good enough to copy for me the wording of a plaque in his lighthouse recording the change in site.
[11] 1825, L.B., 93–5.
[12] A/B 1791.
[13] Most of the records of Trinity House were destroyed by enemy action in 1939–45, and there is little at Holkham until the series of letter-books begins in 1816.

English coast. Between the hours of 2 and 4 a.m. he was shocked to find that the light on Dungeness was so bad that at times neither he nor the ship's captain could see it at all. He complained to Trinity House who passed on the complaint to Coke. Coke apologized and the supervisor of the light, Benjamin Cobb, was ordered to report. Cobb replied that the keeper was certain that the light was as brilliant as any light could be, on the night of Popplewell's complaint; the keeper always 'set his alarum for two, at which hour he never fails going to his lamps'. But Popplewell insisted. The ship he was on, he declared, was so near the point (under low sail, waiting for the pilot boat) that he could easily observe the beach, but 'alas! no light'. Trinity House thereupon called for the lightkeeper to be reprimanded. In reply, the supervisor wrote from Lydd to say that no one else had ever complained. But it soon emerged that all was not well; the lightkeeper himself was 78 and his wife 77 and the lightkeeper had lately been ill and unable to attend to his duties. The old man had his son to help him, but evidently the son was unreliable. So the old keeper was pensioned off, a room was found for his wife and himself to live in, they were given £10 a quarter and a child was provided to look after them. Cobb himself took over the maintenance of the light for the time being—apparently without informing Coke until some time after he had begun to do this. He was enthusiastic about his work, suggested various improvements and was anxious to go for a course of lectures at the Bell Rock. He expressed the fear that Coke was unlikely to come to see his lighthouse but hoped that Blaikie might do so. Blaikie replied from Holkham approving inexpensive improvements but vetoing any others and refusing to agree to Cobb's going on a course. Cobb then suggested some repairs, to prevent rain coming in—apparently rain had even put out the lamps in the winter. Blaikie replied leaving it to him. In 1819 Cobb gave up the superintendence at Dungeness, and Blaikie wrote to Jickling, the head agent in London, to ask him to make sure that Cobb had appointed suitable successors, since, as he said, he was unable to go there himself.[14]

This shows the extent to which the ordinary management of lighthouse affairs occupied attention at Holkham. There is no evidence that Coke ever saw his lighthouse, and it is certain that

[14] 1817, L.B., 233-4, 241, 261; 1818, L.B., 50, 87, 88, 147, 191-2; 1819, L.B., 118, 202, 211.

Blaikie did not. Money came safely in from the tolls without any need for either of them to go to Dungeness; Jickling and his subordinates were left to look after affairs more or less as they pleased —with Blaikie keeping a restraining hand on anything that might involve extra expense, and very occasionally intervening to point out that the lightkeepers held a trust of the highest public importance or that it was essential to avoid provoking damaging reports to Trinity House on the conduct of the light.[15] In December 1821 the lighthouse was struck by lightning and split from top to bottom. Blaikie at once ordered Jickling to inform Trinity House and to ask them what action should be taken; he added that he left it to Jickling in London, and Terry, the new supervisor at Lydd, to do whatever was required. The lighthouse was reinstated at a total cost of £550. 5s. 11d. (with a further estimate of £250 for painting and for a new roof).[16]

Coke and Blaikie showed constant anxiety to win the favour of Trinity House; in 1826, when Trinity House suggested that the lighthouse should be painted red, Blaikie instantly agreed.[17] But Trinity House regarded private owners of lights with considerable distaste. It seemed to them that private owners had no useful functions of any kind, that they did nothing Trinity House could not do—and, in effect, indeed, they were merely grotesquely overpaid agents of Trinity House, who carried out, it is true, the orders of the Brethren, but who introduced administrative complexities which direct management by the Corporation would avoid, and who diverted money paid by shipowners to swelling the incomes of rich landlords, money which Trinity House could put to what it conceived to be better uses—maintenance of navigation marks or charity to distressed sailors.

Private lighthouse-owners were looked on with hostility by another class of persons: those who disliked the exaction from the public of funds to be diverted to private uses; their spokesman was Joseph Hume, the radical opponent of every form of wasteful governmental spending and unnecessary public burdens. A third, more self-interested group, the shipowners, believed that tolls should be levied only as required to maintain navigation marks and lights. Neither the radicals nor the shipowners felt much enthusiasm for Trinity House, which itself extracted larger dues than were

[15] 1820, L.B., 22, 115.   [16] 1821, L.B., 120-2; 1822, L.B., 1, 63-4.
[17] 1826, L.B., 54-5.

required to maintain its services, and devoted the surplus to charities which the shipowners and the radicals regarded as either unnecessary or better provided for in some other way. Still, they recognized the usefulness of the skill and knowledge of Trinity House and that the actual management of navigation marks must remain with it.

Political opinion, therefore, was becoming decidedly hostile to the idea that anyone should derive an income from tolls charged on shipping for the upkeep of lights; tolls should be sufficient to pay for lights and navigation marks and no more. Thus, as the date of the expiry of Coke's lease of Dungeness approached, renewal seemed far from certain. In its Third Report, the Select Committee of the House of Commons on Foreign Trade of 1822, declared, referring to private lights, that it had 'no reason to believe that the Income does not exceed the necessary Expenditure . . . and that great Incomes are not derived from them, and enjoyed by individuals, at the expense of the Shipping of the Country'. It suggested that the grants to individuals should not be renewed when they expired and that the lights should then be handed over to Trinity House.

There followed a clash between old notions of property and new ideas of utility, between Coke's conviction of the sanctity of his right to part of his family inheritance and the contrary belief that he had no right to receive an income from the public without rendering corresponding services. The ultimate outcome was inevitable (even Coke himself sometimes seemed to doubt the validity of his own claims), but it was delayed surprisingly long.

Coke's lease from the Crown was due to expire in 1828. Hanrott and Blaikie discussed the chances of renewal in the summer of 1825. Blaikie regretted that they had not applied for a new lease immediately after the thunderstorm in December 1821. Unfortunately, at the time, Coke 'would not hear of it', but Blaikie thought there was still a chance of renewal, though perhaps with reduced tolls. Blaikie told Hanrott that Coke now thought it might be wise not to mention how much it had cost to build the new lighthouse (the figure would be embarrassingly small). A memorial to the Treasury for a new grant was submitted. It mentioned that the lighthouse had been in the family ever since Richard Tufton was given it by Charles II, that a new lighthouse had been built in 1791, that Trinity House had declared the new lighthouse to be a very good one, and that the lighthouse had been repaired in 1821. Word came from the Treasury at the end of July that Lord Liverpool was examining the memorandum; then it was passed to Trinity House for their comments. Trinity House, as Blaikie told Hanrott, petitioned for the light in 'a document of great length, drawn with great ingenuity, and the Grounds extremely plausible, Reduction of Dues, Improvement of the Navy, and great advantage to the Mercantile Interest, and to the Public Service . . . A most serious loss indeed it will be to Mr. Coke, and how to be met in *ways and means*, I cannot possibly calculate upon.'[18]

In April 1826 the Treasury made known their refusal to renew the lease, a decision 'in conformity . . . with the opinion expressed by the Committee of the House of Commons upon Foreign Trade'. But Coke and his advisers did not abandon the struggle. In June Hanrott called on the Chancellor of the Exchequer, but the Chancellor urged, as Hanrott told Coke, that 'duties payable by the Public ought not to go into the hands of an Individual, as a matter of Profit, and that if under a Grant to the Trinity House a beneficial surplus should arise, that surplus would be employed in charitable purposes, and not increase their funds for their private benefit'. However, Robinson said that the Treasury would receive any further representation that Coke might care to make. Thus encouraged, Coke dispatched to the Treasury an impassioned plea for a renewal of his lease. It was a question of the right of property: 'The Light House and Tolls have been possessed by my ancestors and Myself for upwards of 150 years . . . This property has in consequence been considered a family property and constantly been treated as such from time to time in family arrangements, it having always been contemplated that a renewal would be granted.' It was also a question of a just reward to a devoted servant:

> The Grantee must surely be considered as a Public officer, invested with a trust of the highest Importance, and in consequence with heavy responsibility attached to him. His duties are constant and unremitting, the Intermission of lighting even for a single Night might be productive of very disastrous Effects, and in such an Event, he might be called upon for Compensation to an Amount far exceeding any profits to be derived from the Tolls.[19]

In the end, it seems to have been the firmness of Lord Braybroke, whose lease of other lighthouse tolls had just expired, rather than

[18] 1825, L.B., 77, 80, 93-5, 99, 119; 1826, L.B., 17.
[19] 1826, L.B., 89-90, 96-8.

persuasive rhetoric from Holkham, that won the day. Braybroke made it plain that he would flatly refuse to hand over his lighthouses or the land on which they stood, at Winterton and at Orford, without compensation for the loss of a very valuable part of his settled estate. The government climbed down in 1828 in face of this threat, and Goulburn, the Chancellor of the Exchequer, negotiated an agreement with Braybroke by which he would have one-half of the existing rate of tolls for twenty-one years from 1828 if he promised to convey the relevant land and buildings to the Crown at the end of that period. There was no reason for not giving Coke a new lease on similar conditions and this was done—for twenty-one years from 24 June 1828.[20]

This bargain was by no means good enough for those who felt like Hume, and they kept alive public concern about private lighthouses. In the spring of 1829 the House of Commons called for publication of the facts and figures relating to lighthouses, with special reference to the newly granted leases of private lights. Some of the correspondence relating to the new grants was given, and a statement of the net amount the private lessees had received from tolls for the years 1823-7.[21] Once facts such as these were made public, it was hardly possible for the private lighthouses to survive much longer.

The struggles over the Reform Bill delayed the end of private lights; even so, someone found time in the summer of 1831 to organize at Lydd the signing of a petition to the House of Commons against the appropriation to a private individual of the dues charged for the maintenance of the nearby light. The petitioners were not aware 'that the present Grantee T. W. Coke Esq . . . has by any Public Service entitled himself to the grant of Public Property'. As the manager of the lighthouse commented, 'There seems to be no end to the spirit of hostility which busies itself with the concerns of this unfortunate Light house'.[22]

Hume returned to this campaign on 21 February 1833. He told the House of Commons that he first raised the question in 1821. This is interesting: in that year, Hume had been one of the principal guests at Coke's last sheep-shearing. Hume's zeal for reform was not blunted by his friendship with Coke; neither, more strikingly,

[20] App. (H) and (I) to *Report from Select Committee on Lighthouses* (Parl. Papers 1834 (H.C. 590), XII).
[21] Parl. Papers 1829 (H.C. 241), XXI, pp. 159-61, 178-9.   [22] 1831, L.B., 58.

General Finance, 1822-1842

was Coke's zeal for reform blunted by the fact that the advance of radical feeling jeopardized a substantial part of his income. Hume went on to complain that the government of 1828 had ignored the recommendation of the Select Committee on Foreign Trade that the grants of private lights should not be renewed when they expired. He moved for returns of the amount of duties collected from shipping and of the use to which these moneys were put. For the government (of which Coke was an ardent supporter), Poulett Thomson agreed with the principle of Hume's remarks, and it was Sir Robert Peel who spoke in defence of the leases and the lessees. As he very reasonably insisted, Wellington's renewal of the leases to Braybroke and Coke in 1828 was not a piece of jobbery, for 'Mr. Coke of Norfolk' and Lord Braybroke 'had always been in habits of the most active opposition to Ministers'. Peel went on to argue that the Dungeness lighthouse had been originally built by an ancestor of Coke's 'at his own expence and upon his own ground. If then, the property was his, and it was built upon his own estate, on what principle could the Government take the Lighthouse without making an adequate compensation to the individual?'—evidently Peel thought of the renewal of the lease as a compensation for Coke's pledge to surrender all his property in Dungeness at the end of the 21-year period.[23] Hume, by contrast, looked upon the leases simply as affording means of 'plundering the public' and, according to Lord Braybroke, he once even declared that whoever had advised their renewal should be impeached and that his offence was almost high treason.[24]

Coke was not at his best when meeting hostile criticism of himself; Hume's remarks had struck home. On 20 April 1833, at the St. Andrews Hall in Norwich, nearly 500 people assembled at 5 p.m. to dine in order to mark Coke's retirement as a representative of the County of Norfolk. Coke's old friend, the duke of Sussex, was in the chair. After various Whiggish toasts, the duke proposed a toast to Coke, alluding to his virtues and his work ('the estate was little short of a rabbit warren . . . his estate . . . not worth more than one shilling and sixpence per acre . . .') and Coke, already an almost legendary old man, then nearly 79 years old, retiring from the House of Commons fifty-seven years after his first election as a member, made a long speech in reply. He denounced Tories and

[23] *Parliamentary Debates*, 3rd Ser., XV, cols. 1067-75.
[24] Ibid., XXXI, cols. 166-70; 1835, L.B., 26.

praised Charles James Fox; he told his favourite story, 'In the first leases', one report of his speech runs, 'the Holkham estates were let at 1/6 per acre—the second at 3/- and when they were out, and he offered the farms at 5/- an acre, they were refused. This circumstance induced him first to turn his mind to the pursuit of agriculture . . .'; and, towards the end of his discourse, he turned to deal with Hume:

> There is one point more to which I wish to allude. I have never received a farthing of the public money; my hands are clean. About two years ago, a person of the name of Newton charged me with receiving large sums for Dungeness lighthouse; this I scorned to notice. But the other night in the House of Commons, Mr. Hume alluded to that lighthouse. Now Dungeness is my own private property, quite as much as that of any other person. In the year 1828, the lease expired. The Trinity house wished to have it, but the Government renewed the lease to me, and I aver to you it was the better for the public that they should do so. The offer was made by Government to me—I did not seek it—I made no conditions, and Mr. Peel [sic] the other night did me the justice to say in the House, the Government offered it to me on condition that I would give up the property in twenty years; I did so, and it was a good bargain for Government. Mr. Hume on the evening alluded to, said that I received £8,000 from Dungeness and £4,000 from Harwich, whereas the real fact is that all I receive is £2,000 from Dungeness. Therefore you see how correct this muddle-headed fellow is in the calculations that he is always making, and I can but be glad of this opportunity of stating the truth in this matter . . .[25]

Hume was wrong in suggesting that Coke had ever drawn money from tolls for Harwich and it is strange that he did so. But, for the rest, this passage in Coke's speech was misleading. It is perfectly possible that Coke himself found it difficult to remember how much he received from the lighthouse—he was rich enough, as some of Blaikie's comments show, to take money for granted, as something that was automatically available, and it is possible that he never troubled to analyse his own finances in any detail.

The 'person of the name of Newton' sent a report of Coke's speech to Hume. The latter was by now equipped with the returns he had

---

[25] *Norwich Mercury*, 20 April 1833. The paper preceded its report of the dinner by its own recital of the usual myths: 'the whole of the district round Holkham then produced only a little rye' and so on. *The East Anglian; or Norfolk, Suffolk, Cambridgeshire, Norwich, Lynn and Yarmouth Herald* report (16 Apr. 1833) confirms this report though its wording is often different.

demanded—they included the gross and net revenues from Dungeness for the years 1828-31.[26] With these and the previous returns, Hume was able to compose a crushing reply to Coke's speech, which he sent to Newton who promptly passed it on for publication in the *Norfolk Chronicle*.[27]

Hume pressed forward the year after. In the rather bitter words of Coke's lighthouse agent, 'At the opening of the Session, Mr. Hume moved, as usual, for a Committee on the subject of Lighthouses'.[28] This committee, with wide terms of reference, was appointed on 13 February 1834; Hume was its chairman. It reported in August 1834,[29] recommending, notably, that all lighthouses should be put under the management of Trinity House.

Thus reinforced, Hume moved in March 1835 for leave to bring a bill to 'facilitate the consolidation of lighthouses under one management'.[30] Lord Braybroke was anxious that Coke should communicate with his 'Friends in London', but Coke took up an Olympian attitude, feeling, apparently, certain that his Whig friends in office would spontaneously care for his interests: 'He feels confident that Government will Act with him, as with the other Lessees' as Baker, the Holkham steward, told Coke's solicitor in London.[31] In the next year—the Bill was finally withdrawn in the 1835 session, but reintroduced by Hume in February 1836— when Trinity House in February offered to buy Coke's interest, Coke wrote to its secretary,

> I request you will have the goodness to acquaint the Elder Brethren that I have not had in contemplation to alienate a possession which has been vested in my Family upwards of a Century and a half and that I have not at present any Intention of doing so . . . I rest in perfect confidence, that rights emanating from and guaranteed by the Crown will be duly respected and that if I shall be called upon to surrender this Portion of my Property for the attainment of any Public object, the sacrifice will be requited by a full and satisfactory compensation.[32]

[26] Parl. Papers 1833 (H.C. 170), XXXIII, p. 3.
[27] 20 Apr. 1833.
[28] 1834, L.B., 30.
[29] 1834 H.C. 590.
[30] *Parliamentary Debates*, 3rd Ser., XXVII, cols. 246-55.
[31] 1835, L.B., 26, 55. Baker supposed that 'Mr. Hume must have some hidden object in view, but whether he is connected with the Trinity Corporation or not, I do not know, I can only say it savours much of it, otherwise why should he be so anxious of vesting the whole power of the Light Houses in that Corporation?' 1836, L.B., 10.
[32] 1836, L.B., 13.

By then, indeed, the only question remaining was the size of the compensation. Hume withdrew his Bill in return for a pledge from the government that it would introduce a similar one. This was passed in 1836.[33] It empowered Trinity House to buy out the lessees; if the parties could not agree, one side could refer the question of the prices to a jury, chosen from the special jury list, whose award would be binding. Coke accepted an offer from Trinity House of £20,954. 2s. 5d. for the whole of his interest in Dungeness.

The final irritation to Coke came when Trinity House, correctly, refused to pay the capital to him on the grounds that the lighthouse, land, and tolls had been part of the settled estates. Coke received the interest; the capital was payable only after his death.[34]

iii. *The crisis of 1822 and the recovery*

In 1822 a serious financial crisis appeared to be about to overwhelm the Coke estates. Three separate sub-crises merged, as it were, into one big crisis. One alarming fact was that Coke's annual outgoings could easily be shown to be substantially larger than his income was likely to be in the succeeding few years. The second and third crises depended on the fact that Coke might die quite soon; 1822 was the year of his sixty-eighth birthday, and no one could have expected him to flourish, as he did, for another twenty years. Coke's new wife became pregnant soon after the marriage in February 1822 and the child, a son, was born in December. The possibility of a male heir drew attention to the consequences which the bland disregard of the sinking fund provisions of Lord Leicester's will might produce when Coke died. The whole settled estate might then fall into the hands of Chancery until the provisions of Lord Leicester's settlement were fulfilled. The third crisis was that the proposed provisions of Coke's will would involve, if he left a male heir, a burden on the estates greater than they could bear. Hanrott, the London solicitor, and the devoted Blaikie contemplated the scene with the greatest dismay.

The first crisis, of spending compared with income, was precipitated by two events. One was the agricultural depression which seemed, in 1822, likely to reduce permanently the income Coke derived from his estates. The other was the possibility that the

[33] 6 & 7 Wm. IV, c. 79. [34] 1836, L.B., 129; 1837, L.B., 37.

large income from the tolls for the Dungeness lighthouse would come to an end when the existing lease expired. Blaikie summed the matter up in one of his cries of alarm that filled the year 1822:

> Mr. Coke's income is, of course, rapidly decreasing, his expenditure is also decreasing, but in a very small degree, and by no means in proportion to the falling off in the ways and means. Nor do I expect that the two will ever be brought upon a par; and for this substantial reason. Mr. Coke's benevolent mind outstrips his resources—It is a virtue in him carried to excess, and for which I see no remedy in this world.[35]

In October 1822 Blaikie drew up for Hanrott a statement of Coke's income and outgoings based on the position in 1821. First, he gave the net yearly income from all the estates, including Dungeness, as £34,262. 7s. 9d., a figure reached after deducting all expenses of management, repair bills, and taxation. He went on to calculate the outgoings this income had to meet or might have to meet in the future. In the first place came the large sum to be paid out for interest on debt and for annuities, together with the small allowance—£240 a year—for renewals of College leases. The total was no less than £14,468. 19s. 4d.; of this £2,835. 16s. 8d. was interest on the debts on the settled estate—the burden Coke inherited from his great-uncle, the builder of Holkham. Other elements in this interest charge were hypothetical—the annuities would only come into operation after Coke's death, and the yearly interest of £1,100 on Coke's daughter's portion was never required since her £22,000 was somehow found in 1822—perhaps from the portion Coke must have received when he himself was married earlier in the year.[36] These hypothetical items made up £3,815. 5s. 0d. of Blaikie's total. Next, Blaikie enumerated other outgoings on the upkeep of the estates—mostly for estate repairs paid for by the Holkham office. The total was £3,012. 4s. 9d. Then there came the cost of sustaining Coke's way of life as a great landowner, the 'expenses considered as belonging to the domestic establishment'. These reached £11,939. 12s. 7d., to which Blaikie added £4,000 for Coke's personal spending.[37] The total expenditure on personal

[35] 1822, L.B., 101. Blaikie's underlining.
[36] 'Another ebullition of happiness from my father this morning on my prospects, and self-congratulation at having paid my fortune which in his own words, if the corn laws are repealed, he "might not have been able to do three years hence".' Elizabeth Coke to John Spencer-Stanhope, 10 Nov. 1822. A. M. W. Stirling, *The Letter Bag of Lady Elizabeth Spencer-Stanhope*, ii. 50.
[37] See Appendix 4 for details.

consumption was thus £15,939. 12s. 7d., which may be compared with the £7,904. 2s. 3½d. devoted to the same purposes precisely a century before. (The greater part of Coke's spending was on his life in Norfolk.) Blaikie reached a grand total for Coke's annual outgoings of £33,420. 16s. 8d. This was all very well—it left a credit balance of £841. 11s. 1d., but further calculations darkened the picture.

In the first place, decay of agricultural prosperity could be expected to bring about a permanent fall in rents—Blaikie thought 10 per cent off gross rental would be a modest estimate, and that would take away £4,700 a year. Blaikie noted, however, that a reduction of one-half per cent in the bond and mortgage debts would bring a fall in outgoings of £215 a year. He subtracted this, reducing the expected decline in income to £4,485 and he set this against the existing surplus of £841. 11s. 1d., making a (slightly miscalculated) 'deficiency of annual Income in future' of £3,641. 11s. 3d. This was bad enough, but there was worse to come. The loss of the lighthouse might be expected in five years or so and Coke, or his heir, would then lose another £5,700 a year. What is more, Coke's nephew might marry and Coke had promised to increase his allowance by £1,200 a year if he did. Thus a total deficiency could be predicted of the frightening amount of £10,541. 11s. 3d. per annum. Even if the hypothetical elements had been removed (Blaikie evidently wished to paint a black picture) and the nephew's allowance and the hypothetical interest charge taken away, the deficiency would still have been £5,526. 6s. 3d.[38]

The second crisis is best stated in Blaikie's words:[39] 'And you are also aware that in the event of a minority, the Infant will become a Ward of Chancery, and that the Chancellor will claim upon Mr. Coke's private Estate, the Amount of £56,716. 18s. 10d. mortgage upon the settled estate which ought to have been paid off according to the provision of Lord Leicester's Will.' It will be recalled that the builder of Holkham had prescribed that £3,000 a year of the income from the estate he settled by his will of 1756 should be devoted to paying off his debts, all of which should therefore have been cleared off by 1805 or so, if Coke had not brought all such repayments to an end after 1789.

[38] In H.F.D., 112. Statement showing the annual Income and Expenditure of Thos. Wm. Coke Esqr.
[39] To Hanrott, 30 Dec. 1822. 1822, L.B., 177.

The third crisis was partly caused by Coke's own proposals of 1822. The effect of the provisions of the will he wished to execute would be that, at his death, his private estate—the non-settled estate—would have more charges laid on it than it could bear. This was a result of the remarkably restrictive provisions of Lord Leicester's will, under which nothing could be charged on the estates he settled except an annual sum for jointure for wives of tenants for life. Everything else had to be provided from the unsettled estate. In October 1822 Blaikie produced a 'Statement showing the amount of Debts, Charges, Legacies and Annuities to be provided for out of the Estate of Thos. Wm. Coke Esq, and the Funds to meet the same'.[40] Debts and charges, actual and prospective, were £276,274. 12s. 2d. The prospective charges were for the portions of children of Coke's brother, which Coke had undertaken responsibility for, and the portions for the potential younger children of Coke's new marriage. Coke's proposed legacies would add another £20,860, making a total of £297,134. 12s. 2d. He further proposed to bequeath life annuities, which, added to annuities already being paid from the profits of the estate, would make a total of £2,931 per annum after Coke's death. To meet all this, Coke's unsettled real estate Blaikie put at a capital value of £319,200, and estimated his personalty at £29,410. 0s. 0d. (of which a mere £1,410 was in securities due to Coke), a total capital of £348,610. A capital balance of £51,475. 7s. 10d. would therefore remain to deal with the annuities. But 5 per cent of this would only be £2,573. 15s. 4d., and a deficit on the life annuities of £357. 4s. 8d. would emerge.[41]

It seems to have been difficult to persuade Coke himself to take all this seriously. In September 1822 Blaikie told Hanrott,

I do assure you, I am quite appalled with the present and future prospect of Mr. Coke's affairs. Would to God Mr. Coke could but see those matters in the serious light that I do (and my opinion is not taken up hastily). He would then take immediate steps for averting the calamity which must otherwise inevitably befall him and his family; relieve himself from much present inquietude on pecuniary matters, and insure his future comfort in that very important point.[42]

To meet the crisis, Blaikie and Hanrott soon agreed to employ the only possible means—a sale of land. 'It is submitted', Hanrott

[40] In H.F.D., 112.
[41] Again, Blaikie's calculation was falsified by Coke's success in paying Elizabeth's portion at the end of 1822.   [42] 1822, L.B., 126-7.

wrote, 'that the best course to adopt will be to take immediate steps for the sale of the Estate in the County of Bucks. The doing so will enable the payment of a very considerable proportion of the above debts and charges in Mr. Coke's lifetime; and effect an important relief to him in many respects.'[43]

The sale was not finally negotiated until September 1823, when an agreement was made with John Farquhar for the sale of Hillesden, Buckinghamshire, for £127,000. This estate comprised 3,105 acres. Its 'full rental' was £4,509; its 'reduced rental' was £4,009. It was subject to out-payments of £471 a year.[44] The advantages of the sale are easily demonstrated. In return for a loss of income of £4,000 a year, Coke would be able to pay off debts of a capital value of £127,000. By doing so, he would save interest payments of, at 5 per cent, £6,350 a year, at 4½ per cent, £5,715 a year or at 4 per cent, £5,080 a year. Repaying debt from the proceeds of the sale would therefore give Coke increased disposable income of from £1,080 a year to £2,350 a year.

Hanrott and Blaikie were concerned to prevent Coke's comparative irresponsibility leading to his using the proceeds of the sale as income. Immediately after the sale, Blaikie wrote to Hanrott with evident relief: 'I am happy to inform you that Mr. Coke has stated to me positively that the whole of the purchase money from the Hillesden estate shall be appropriated to the payment of the Bankers, Bonds and other debts bearing high Interest.' Even so, Blaikie continued to show alarm about what Coke might do with the money from the sale. Early in 1824 he wrote to Hanrott:

> Pray do keep a distinct and clear account of this money transaction and do not allow it to be muddled and mixed up with Mr. Coke's other Banker's Accounts—on this important point, <u>The Flag should be nailed to the Mast</u>. If the money is diverted to other purposes, than those contemplated in the sale of the Estate, Mr. Coke's pecuniary resources will be alarmingly narrowed, and the blame will untimately be attached to both you and myself—I have this matter much at heart.[45]

It was done as Blaikie wished and the sale made a major contribution towards improving the condition of Coke's finances. But there were other causes of the excellent condition they attained before Coke's death. There was the reduction in interest rates,

---

[43] In H.F.D., 112—General statement of the property and affairs of T. W. Coke Esq. Blaikie had suggested such a sale—Blaikie to Hanrott, 24 June 1823.
[44] 1823, L.B., 180-3.   [45] 1823, L.B., 184; 1824, L.B., 66-7 Blaikie's underlining.

which reduced the cost of servicing debt. Most important of all, it turned out that Blaikie's assumption that rents would permanently fall from the level of 1821 was unfounded. Furthermore, income continued to be derived from the lighthouse until 1835.

Blaikie counted on reduction in interest from 5 per cent to $4\frac{1}{2}$ per cent on debts of £43,000, when he drew up his statement of Coke's income and expenditure in 1822.[46] Early in 1824 Hanrott wrote to Blaikie:

> The time is now arrived when in my opinion, Mr. Coke ought to insist on the reduction of the Interest on all his Mortgages (not for a stipulated period unexpired) to £4 p.Cent, that rate seems quite general and likely to continue until further lessened. In case of refusal, I do think it will be worth while to pay off, for though the expenses of transfer are to be considered, yet they would soon be made up by the interest saved, and I believe I could get the whole money which might be wanted without going out of our own office—I think it likely also that even in those cases where the term fixed for the duration of the Mortgage is not expired, the parties might agree to a reduction.[47]

Hanrott was able to bring down many of the rates of interest on Coke's debts to 4 per cent in 1824. The debts whose interest he was unable to reduce were marked down for early repayment. In 1826 interest rates went up again, to $4\frac{1}{2}$ per cent and then to 5 per cent. But by 1827 so large a quantity of debt had been repaid since 1822 that this rise in interest was not a matter for great alarm.[48] In the 1830s low interest rates returned, and in 1836 Croft, Coke's new London solicitor, wrote that he had given notice to Lady Beauchamp that Coke proposed to reduce to $3\frac{1}{2}$ per cent the interest on the £17,000 due to her on mortgage. Croft secured Coke's agreement that Lady Beauchamp should be paid off if she refused, 'as Money can, if wanted, readily be had in large sums at $3\frac{1}{2}$ p.Cent'. Lady Beauchamp agreed.[49]

But the chief reason why the crisis that threatened in 1822 did not develop was that the farming slump did not persist, so that rents did not fall permanently. As a result, Coke was able in the 1820s to repay large amounts of debt from income, as well as from the proceeds of the sale in Buckinghamshire. His position was further strengthened when it turned out to be possible to retain an income, though a reduced one, from the lighthouse tolls, until the

[46] H.F.D., 112. [47] 1824, L.B., 30.
[48] A/B accounts current, 1822–7. [49] 1836, L.B., 67–70.

year 1836, by which time enough debt had been disposed of to make tolerable the final loss of the lighthouse. What is more, by then, rents were beginning to show a definite upward trend. After 1838 they rose rapidly, and by the time of Coke's death in 1842 the position was excellent.

The effects on Coke's income of the sale of Hillesden, and of the loss of the lighthouse are best illustrated by stating Coke's income from all sources (except the small amount from Lancashire). The figures in Table R are to the nearest pound; net income is the figure recorded in the Audit Books, a figure subject to further deductions for estate purposes; until 1836 some of the fluctuations are due to fluctuations in lighthouse returns.

TABLE R

	Gross	Net
	£	£
1821	47,200	37,357
1822	45,734	42,231
1823	45,953	41,286
1824	41,416	38,010 (sale in Bucks.)
1825	41,860	39,373
1826	41,698	36,824
1827	41,914	35,888
1828	40,787	36,132
1829	41,265	34,044 (net returns from lighthouse halved)
1830	39,211	32,548
1831	39,595	32,344
1832	39,668	32,909
1833	39,711	31,163
1834	40,003	31,529
1835	40,599	31,039
1836	35,661	27,305 (lighthouse lost)
1837	35,930	29,419
1838	36,465	30,236
1839	37,468	32,294
1840	38,814	33,563
1841	38,610	33,591
1842	39,117	31,444

No doubt economizing at Holkham also contributed to the bringing about of a surplus of income over spending, sufficiently large, when combined with the effects of the sale of Hillesden, to make it possible to pay off enough debt to reduce the interest charge and so to increase the surplus again to a point at which the loss of the lighthouse could easily be borne. In January 1822 Blaikie told a stricken tenant that 'Mr. Coke has made and is still making great reductions in his expenditure, so as to endeavour if possible to meet the times'.[50] Blaikie himself took a cut in salary: in 1821 he got £650; in 1822 only £550.[51] It is to be noted that there was no mention in Blaikie's letters to Hanrott of any possible savings to be made in spending on the upkeep of the existing estates. There was, however, one investment which would normally have been made which was not made because of the financial crisis. In August 1827 Coke wrote to refuse Lord Spencer's offer to sell him his estate at North Creake. This was a substantial property, of more than 1,000 acres, adjoining Holkham. 'In order to disencumber my property . . . I was induced a short time since to dispose of my Buckinghamshire property . . . desirable as the purchase of your Lordship's Estate would be from its vicinity to Holkham I dare not venture upon it.'[52]

By 1827, however, the threatened crisis could be seen to have been overcome. On 24 January 1827 Blaikie sent to Hanrott a statement of Coke's debts, together with an account of his capital transactions since 1816—the year in which Blaikie came to Holkham. The scene was transformed. Coke's total debts were now only £84,000 (excluding a mortgage on the settled estates of £11,100 which was about to be discharged). Only £10,000 remained of the mortgate debts left by the builder of Holkham on the settled estate, £45,000 was due on mortgage on the Derbyshire estate and £29,000 on bond. Coke had even lent £2,000 to one of his daughters, and another £710 was due to him on various securities. In addition, he had 'sundry shares', apparently of no great value, in 'the London University, the Cross Keys Wash Bridge, Norwich Union Fire Office, Norwich Literary Institution and other Securities on which Instalments have been paid, but not paid in full—There is also a share of £300 in Drury Lane Theatre, which has been paid in full, but no Interest receivable at present.' Since 1816 debts to the value of no less than £165,431. 4s. 11d. had been paid off. In the same years, £12,662. 5s. 0d. had been laid out on the purchase of land. The Buckinghamshire sale produced £125,200 and the sale of the remnant of the Minster Lovell estate in 1824 brought in £3,500.[53]

[50] 1822, L.B., 15.
[51] A/B 1821, 1822 (accounts current).
[52] Althorp MSS. Corr., 2nd earl Spencer, Box 147.
[53] Sold to W. E. Taunton. 1824, L.B., 140-1.

Setting off purchases and money lent against sales, and deducting the balance from the total of debts repaid, Blaikie concluded that £52,103. 10s. 4d. had been saved from income and applied to the discharge of debt since he came to Holkham in 1816. About £21,000 of this must have been saved out of income since October 1822, when Blaikie worked out his gloomy statement. Blaikie was justifiably pleased. 'You will observe', he wrote to Hanrott, 'that the Money paid in discharge of incumbrances and for Estates purchased of late years amounts to a very considerable sum, probably much larger than you had anticipated. It certainly far exceeds any previous calculation and shews the husbandry of our resources in a far more favourable light than any previous statement of a similar nature.'[54] Hanrott was equally gratified: 'I am delighted with your statement. I was confident that a very large sum had been paid and that much had been done, but was not aware to such an amount as your statement shows, but which appears to me quite correct.[55]

Thereafter no other threats to Coke's prosperity appeared. His financial security was made complete when a renewal of the lighthouse grant was agreed to in May 1828. Repayment of debt continued: in January 1828 a £10,000 mortgage was repaid—this was the last of the debts left by Lord Leicester in 1759. Blaikie expressed his satisfaction, but Hanrott replied 'I should like to live to see not a sixpence owing in any direction' and Blaikie then reverted to his habitual gloom: 'That desirable object I despair of ever seeing accomplished. Indeed, there can be no manner of doubt that, if Mr. Coke loses the Light House, and continues the present rate of expenditure, he must either allow a portion of His Bills to remain unpaid or further encumber his Estate.'[56] News of the renewal of the lighthouse grant soon come to relieve him of his probably somewhat exaggerated fears.[57] In February 1829 another mortgage was paid off: £12,000 to E. P. Lygon—this mortgage had been on the unsettled Derbyshire estates.[58] Only one mortgage debt remained—the £17,000 due to Lady Beauchamp, which was not disposed of until the spring of 1842.[59]

[54] 1827, L.B., 31-3.   [55] Ibid., 38.   [56] 1828, L.B., 8, 17.   [57] Ibid., 52.
[58] It seems to have been no accident that these loans were repaid in January and February—the end of the estate financial year. Blaikie wrote to Hanrott in July 1830: 'Mr. Coke . . . is desirous of paying off Col. Lygon's Mortgage of £16,000 . . . and thinks He will be enabled to do so, without much inconvenience, after closing the year's Rent account in Jany. next year.' 1830, L.B., 70.
[59] 1841, L.B., 79.

General Finance, 1822-1842

The bond debts due in 1827 were £24,000 to Coke's brother, Edward, and £5,000 to Overman's executors. The £5,000 seems to have been paid before 1842. The £24,000 would have been paid off if payment of the compensation money for the lighthouse tolls had not been delayed.[60] It was only possible to pay £4,000 in 1837 and another £8,000 in 1838.[61] Another £4,000 must have been paid some time before 1842, for at Coke's death only £8,000 remained due on bond, now owing to T. W. Coke, Coke's nephew.[62] This £8,000 was all that remained, when Coke died, of the vast debts of 1822.

Thus Hanrott's dream of 'not a sixpence owing in any direction' was nearly fulfilled, largely as a result of the actions of Blaikie and Hanrott after the war, which prevented Coke from rolling down a slope from increasing debt and increasing interest payments to further debts and still larger payments.

Lord Leicester died on 30 June 1842, at the age of 88, having been in control of the estates for sixty-six years. The personal estate left by Lord Leicester was small, apart from the value of rents falling due, farming stock and crops, furniture, pictures, and so on. After the loss of the lighthouse, his income had been derived almost exclusively from land—his other capital was very small. He had nothing at the bank and only £9 cash in his house; debts due to him and the securities he held were worth only about £1,200, including a share of £100 in the Lynn theatre, £100 in the Agecroft turnpike trust, £150 debenture in the Reform Club, three shares in the Drury Lane Theatre, £200 on a mortgage of the workhouse at Bawdeswick, and £250 due to him on note of hand. He had made no investments except in his own estates.[63]

After the 1822 crisis, though, capital was flowing from Coke to the outside world, but it was used to repay debt. After the Hillesden sale, however, it is probable that Coke maintained, on average, credit balances with his bankers. Income from rents was accumulated over short periods for use in repayment of debt, and sometimes put temporarily into government securities. In 1824 Blaikie reminded Hanrott that they had agreed to use the money from the sale of Hillesden for 'clearing the Bankers Books (Gurneys of

[60] 1836, L.B., 78.
[61] 1844, L.B.(B), 3.
[62] A/B 1845. Statement of Legacies and Interest paid since the death of the earl of Leicester.
[63] H.F.D., 124 and 131.

Norwich will require upwards of £30,000 . . .)'.[64] In February 1827 Blaikie asked Hanrott to give him 'due notice of the precise day when you will want the £11,100 to pay Miss Herbert. Mr. Coke will then give an order to Messrs Hammersleys to sell the remainder of the Exchequer Bills, and an order to Messrs Gurney to remit the balance including interest to Messrs Hammersleys.'[65]

In spite of the burden of debt he had inherited, in spite of his own extravagance and his costly electioneering, in spite of agricultural crises and the loss of the lighthouse, Lord Leicester passed on to his son a secure prosperity, solidly based on a rich and flourishing Norfolk estate.

[64] 1824, L.B., 31.
[65] Gurneys were Coke's Norwich bankers. Hammersleys, who transferred their business to Coutts in 1840, were his London bankers. Only a few relevant records of Gurneys survive, only a signature book of Hammersleys. At present (1974) Coutts records are difficult of access. I am grateful to the Head Office and Norwich Local Head Office of Barclays Bank for their help and I must also thank Miss M. V. Stokes, archivist at Coutts.

13

Some conclusions

A STUDY of a single estate cannot establish new historical generalizations. It may, however, weaken or destroy old ones. The history of the Cokes shows that the heroic interpretation of an English eighteenth-century agricultural revolution as a rapid economic transformation accomplished by the example and precept of one or two great men is doubly unsound. First, the impact on agricultural history of one of the heroes, Coke of Holkham, was much less drastic than was claimed; secondly the pace of agricultural change on his estate, an estate agreed by contemporaries to be in the forefront of progress, was slow enough to make questionable the whole notion of an agricultural revolution. Vanity and concern for political popularity caused Coke to accept and encourage exaggeration of his economic beneficence.

The main economic function of the landlord was to accumulate capital from his estate and from outside his estate and to use it to advance agricultural productivity. Under Coke's predecessors this was done more effectively than under Coke himself. In the first half of the eighteenth century capital provided by the landlord helped in the widespread improvement of hitherto uncultivated or lightly cultivated land. It was used for the direct improvement of the soil, in clearing land and in marling and claying as well as in the provision of farm buildings. Under Coke of Norfolk, tenants provided a larger share of the capital used for soil improvement.

The relationship between the Cokes and their tenants, especially in Coke of Norfolk's time, served in other ways to advance agricultural progress. Through husbandry covenants the use of the basic elements in Norfolk husbandry, turnips and sown grasses eaten by sheep and rotated with corn crops, was encouraged or prescribed. Under Coke the clauses of leases became very elaborate and there are signs, especially after Blaikie became steward in 1816, that they were enforced. Thus the farming of tenants advanced towards the celebrated Norfolk four-course rotation, though this was by no means common until the increase, in the early nineteenth century, of the use of non-animal fertilizers. Coke and his stewards, especially Blaikie, also stimulated good husbandry among tenants by exhortation and preaching. Landlord and tenant acted in economic and intellectual collaboration.

Another means of stimulating agricultural production was through example. Coke attached great importance to his own farm as an educational instrument. It seems to have helped, for example, to bring into Norfolk mangel-wurzels and improved breeds of sheep and to have hastened the spread of drilling. There are signs, however, that the farm was as much an uneconomic piece of display on Coke's part as a serious economic unit.

One respect in which evidence from the Coke estates may help to establish a generalization is the fact that the Cokes were borrowers not lenders. They ought to have been more able than many great landowners to avoid borrowing—since they had fewer daughters and younger sons to endow—but their extravagant mode of life dictated it. Their profits did not help the advance of commerce and industry; it was rather the other way round—capital flowed towards the estate, not away from it. In the early eighteenth century capital moved into Coke lands from local people, who were sometimes tenants or servants, or on mortgage security from other landlords or from financiers; later the borrowing from local lenders was replaced by loans from banks.

Some of the capital available to the Cokes was invested in the estate. Land was purchased to add to it, sometimes substantial areas extending the existing limits of Coke ownership, more regularly (and more significantly in increasing agricultural productivity) small areas intermingled with existing Coke properties. Such land was bought whenever opportunity occurred and, in consequence, enclosure and remodelling of field shapes was made practicable. On this estate, enclosure by act of parliament was much less important than redistribution by the landlord after purchases.

As landlords the Cokes fulfilled economic functions. They were eager to maximize their incomes. Their lavish spending, however, and the social and cultural aspirations which it supported placed limits on their enrichment. Building, collecting, and politics all competed for money with productive investments.

After the South Sea crisis no investments were made outside the estates, so that the possibility of higher returns was lost. This was because Coke of Norfolk, like other English landowners, used his estate not only as a money-making asset but as the foundation for the social prestige and political importance in his county which were his main aims. Both these aims required methods of management of his estate in which the search for profit was tempered by the search for deference and esteem, at any rate from the articulate and prosperous sections of rural society. In this way English country society acquired that characteristic cohesion and conservatism which distinguished it from countries where the aspirations of landowners were different.

APPENDIX 1

The management of the lighthouse

By the leases from the Crown, Lord Lovell's officers had the right to a place in every customs house and orders were given to 'Customers, Collectors, Comptrollers, Surveyors, Searchers and Waiters' and all their clerks at all harbours, and to Captains, Lieutenants and other officers, 'Wardens portreeves and keepers' of ports, and all other customs officers, that they should not permit any goods to be discharged or clear any ship without a ticket or note under the handwriting of Lord Lovell's deputies.[1] At each port there were collectors who took the dues. These men were probably employed at the ports as Crown officials and supplemented their pay by acting for the Cokes and, often, other owners as well. They retained a poundage or commission on the amount they collected, the proportion of which varied, being largest where traffic was least. In 1736 the poundage at London was 10 per cent and at Bristol 15 per cent. In most places it was 20 per cent, as at Padstow, Scilly, Penzance, Falmouth, and Liverpool, for instance, and occasionally 25 per cent, as at Dartmouth. The largest amount of dues was collected at London, where, in 1737, over £1,000 was taken. Only one other port exceeded £100—that was Newcastle. At only twelve others was more than £20 taken: at Portsmouth, Deal, Bristol, Southampton, Falmouth, Exeter, Yarmouth, Poole, Shoreham, Lynn, Weymouth, and Hull, in that rather surprising order of importance. It must be remembered that dues were payable only on ships passing the lighthouse, and the figures are a measure of the channel-going shipping using the ports, not of all the shipping there.

At the lighthouse itself there were two lightkeepers: in 1736 John Harden and John Lanes, who were paid a joint salary of £10 a quarter. These two were supervised by a resident of Lydd, in 1736 one Charles Coxsell, who was, to judge by a letter of his, only just literate. He was responsible, too, for ordering fuel—the light then was fed only by coal. The whole organization was controlled by a chief agent, or agents, in London. In 1736 they were Mr. Henry Hargrave and Mr. J. Whormby, who received £100 a year between them for their services and to cover postage and petty expenses. Whormby was an important figure in lighthouse affairs, being secretary to Trinity House from 1729 to 1750. He acted as agent for other private owners of lights and must have gained a substantial income in that way; furthermore, he held shares in some private lights. This sort of dual allegiance would not have been possible

[1] Letters patent of 28 June, 13 Geo. II, in Muniment Room.

in later years. The agents appointed all the men involved in the affairs of the light and through them came communications about the lighthouse to the steward at Holkham. They drew up annual accounts, showing how much money had come in and from what ports, how much had been spent, and for what purposes, and remitted the net revenues. They made whatever directions were necessary in managing the lighthouse. The whole enterprise, therefore, looked at from Holkham, caused gratifyingly little trouble.

APPENDIX 2

A note on the word 'Break'

'BREAK' is an extremely puzzling word. Sometimes it means 'new enclosure', sometimes it refers to enclosures that are perhaps only temporary, sometimes to enclosures that are clearly permanent. Usually it means enclosure from land used mainly for sheep pasture, but in Longham, where it referred to an enclosure from former arable strips it was clearly not used in that sense. The word does not necessarily mean 'enclosure' at all. A map of Tittleshall records, in its margin, a farmer holding 230 a. 3 r. 2 p. of 'Break now Enclos'd'.[1] Sometimes 'Break' is 'Profitable Ground';[2] sometimes it is equivalent to sheep-walk.[3] A clause in a lease agreement of 1696 for Waterden farm uses 'break' as meaning inferior land: the tenant is not 'to plow or sow any of the Arable Land out of Course: nor to plow or sow any of ye Infield land more than five Cropps and the Breckes but three only, before ye same shall be somertilled or laid for pasture'.[4] The rotation dictated for the 'breaks' might be superior, but that is not the point and here certainly break means inferior or less highly cultivated land, as, indeed, one would expect if the word corresponds to any survival of the 'infield–outfield' system of medieval days.[5] Marshall defines the word 'Breck' simply as: 'A large new-made inclosure (a Break)', a definition which doubtless reflects the usage of his time.[6]

[1] Map 74/4.
[2] Longham Map, 5/93.
[3] Norwich P.L. MS. 25 x 1, Flitcham Deeds, 472.
[4] Waterden Deeds, 58a.
[5] See H. C. Darby and J. Saltmarsh, 'The Infield–Outfield system of a Norfolk Manor' in *Economic History* (Supplement to *Economic Journal*), iii (1934–7, article Feb. 1935), 30–44.
[6] W. Marshall, *The Rural Economy of Norfolk* (London, 1787), ii. 376.

APPENDIX 3

Calculations, *c*. 1815, about Longham Hall Farm, probably made by the tenant

Wheat 105 acres:

	Cbs.		Cbs.		£	s.	d.
25 of which will grow	7 per Acre making	175	515 Coombs at 30/ per Cb.			
50 of Do at	5 per Do	250		772	10	0
30 of Heath at	3 per Do	90				

Barley 105 acres:

50 of which	at	6 per Acre	is	300	645 at 12/ per Cb.			
25 of which	at	9 per Acre		225		387	0	0
30 of Heath	at	4 per Do		120				

Turnips	105	
New Layer	105	
Old Do	52	Consum'd by Stock (as below)
Oats	52	
Pasture	70	
	594	of Farm

Return of Stock

	£	s.	d.
30 Beasts (Grazing) to pay £6 per Head	180	0	0
10 Score Hoggetts at £20 per Score	200	0	0
Wool of 10 Score Ewes	50	0	0
	1589	0	0

Outgoings of Farm

	£	s.	d.
Rent of Farm	570	0	0
Poor Rate	80	0	0
Taxes with deducting of Property Tax	30	0	0
Average of Expences of Labour per Week for 48 weeks £10	480	0	0
Expences of Harvest	110	0	0
Servants Wages	68	0	0
Tradesmens Bills for Implements	120	0	0
Seed Wheat and Barley	210	0	0
Grass Seeds &c	90	0	0
Household Expences	250	0	0
Interest of Capital (5000£)	250	0	0
	2258	0	0
	1589	10	0
	668	10	0

APPENDIX 4

From H.F.D., 112, October 1822:
Expenses considered as belonging to the domestic Establishment

Amount paid for Hops and for Excise Duty on Malt—the Barley being comprised in Mr. Bulling's farm account	333	10	0
Household Disbursements by the Cook and Housekeeper (exclusive of those articles furnished from the Farm) amounting in value to £1,585.14.1	1,704	12	7
Wine and Spirits	899	0	0
Housekeeping Bills in London	472	6	9
Servants' wages, travelling expenses, carriage, postage, &c.	2,050	0	4
Servants' liveries, coals, oil, wax and tallow candles, vinegar, linen, turnery, perfumery &c.	2,620	3	5
Oats, horses, saddlery and farriery	712	19	9
Average annual expenditure on furniture, pictures, statuary and china	700	0	0
Books, stationery and newspapers	165	17	4
Assessed taxes, insurance and poors rates	611	13	10
Loose for labour, nurserymen and other bills relating to the gardens, pleasure grounds &c.	803	4	5
Casual and charitable donations from Holkham	360	13	0
Flax dressing schools	85	1	2
Hire of house in town	220	10	0
Estimated annual amount of medical attendance, say,	200	0	0
	£11,939	12	7
Personal expenses taken on an average of 2 years and for any omitted items paid through the London bankers	£4,000	0	0

APPENDIX 5

A note on prices

WITHOUT knowing how prices for farm produce moved in any period, we cannot decide how far increasing incomes from the land were the result of increasing efficiency. Unfortunately, it is remarkably difficult to produce analyses of price movements in an area such as north-west Norfolk, especially for any period before 1771. After 1771 returns were published each week in the *London Gazette* of corn prices in the various counties of England and Wales. These returns were based on prices at various selected market towns within each county. From these returns a national average price for each grain for each week was worked out. In the Board of Agriculture publication of 1902 of *Agricultural Returns*, etc., for 1901 (1902 Cd. 1121) there were published a series of annual average national prices, based on the *Gazette* returns, for wheat, barley, and oats. These national figures[1] must be used with caution: until 1821, they appear to be unweighted averages—the prices at a market with a smaller turnover would rank equally with prices from a larger corn-dealing centre; prices from a county of small arable culture equally with prices from Norfolk, for instance; prices from weeks of low volume of sales equally with prices from weeks of large trading. Nevertheless, they are good enough to indicate a general trend. The graph shows five-year moving average prices for wheat and barley over the years 1771–1844, together with prices for individual years, based on the Board of Agriculture publication of 1902. Of course, prices in Norfolk were not the same as average national prices. The Table on p. 208 indicates this difference: prices are *London Gazette* averages per quarter of 8 bushels (equal to 2 coombs). National prices come from Cd. 1121; Norfolk prices for 1771, 1773, and 1815 were worked out directly from the average Norfolk prices given each week in the *London Gazette*; those for 1787, 1802, and 1807 are from the summaries of *Gazette* returns in the *Annals of Agriculture*. The Norfolk prices have been converted from Winchester bushel units to Imperial bushels.

Norfolk prices seem, therefore, to have been consistently below the average price elsewhere, though not in a constant proportion. Probably the disparity diminished as transport improved later on.

For the years before 1771 there are no figures in print that are useful for the present purpose. There are few respectable barley prices at all

[1] Figures for coastal counties, especially for Essex, London, and Kent, were more carefully worked out.

Appendix 5

	Barley				Wheat			
	Norfolk		National		Norfolk		National	
	s.	d.	s.	d.	s.	d.	s.	d.
1771	22	3	26	5	44	11	48	7
1773	23	8	29	2	52	1	52	7
1787	21	3	23	4	38	7	42	5
1802	31	0	33	4	67	4	69	10
1807	36	1	39	4	68	5	75	4
1815	25	9	30	3	58	11	65	7

and barley was the most important crop in north-western Norfolk: the earlier the years under examination, within the eighteenth century, at least, the greater its importance. Series of wheat prices are, therefore, of doubtful value; series of wheat prices for other parts of England, are, in the absence of a reasonable mechanism for providing comparable Norfolk figures to test the extent of the disparity between them and north-western Norfolk prices, entirely useless.

The ideal would be figures showing what tenant farmers actually got for what they sold; the second best would be figures of what was got for the produce of the farm in hand at Holkham. Unfortunately, no continuous series of Holkham prices can be found, but there are enough to give an extremely useful guide to probable movements in the prices north-western Norfolk farmers could obtain for their produce, especially for their corn—the figures for livestock products are more unsatisfactory. Agricultural prices are difficult to interpret. Short-run changes due to contrasts between good and bad harvests make it difficult to detect long-term trends. Different qualities of produce command different prices, so that two transactions in the same commodity are not necessarily comparable: in February 1817, as an extreme instance, the Park farm sold 88 coombs of the best malting barley at 32s. per coomb, while in the same month, 54 coombs of inferior barley were sold at only 13s. A local difficulty in using figures from the Holkham farm is that many of its sales were internal—made to other departments of Coke administration, sales whose nominal price might or might not accurately reflect the state of the market. Here are the prices at which barley, wheat, and rye were sold from the Holkham farm in various years from 1724 to 1826. Except for the years in the group 1738-46, they exclude internal prices and are the prices at which genuine commercial transactions took place (they are in shillings per coomb); not all the transactions, especially in the earlier years, certainly took place in the years given as distinct from the immediately preceding or succeeding months.

Appendix 5

	Wheat	Barley	Rye	
1724	12s. 0d.: 16s. 0d.	8s. 0d.: 10s. 0d.		Account of the
1725	16s. 6d.:	7s. 6d.: 8s. 0d.	11s. 6d.: 12s. 0d.	Domestick
1726	15s. 0d.: 19s. 6d.	7s. 0d.: 8s. 6d.	10s. 6d.	Disbursements
1727	18s. 0d.: 20s. 0d.	10s. 6d.		in Norfolk,
1728	20s. 0d.	10s. 0d.	12s. 0d.: 13s. 3d.	1724-28
1732	9s. 6d.: 10s. 0d.	5s. 0d.	6s. 0d.	Thorold Rogers,
1733	11s. 6d.: 12s. 0d.	6s. 0d.: 7s. 0d.	8s. 0d.: 9s. 0d.	*Agriculture and*
1734	16s. 0d.: 16s. 8d.	5s. 0d.: 6s. 0d.	9s. 0d.	*Prices*, vii. 636-
1735	15s. 0d.	6s. 0d.		704. Holkham
1736	13s. 0d.: 14s. 0d.	8s. 6d.	9s. 0d.: 10s. 0d.	Deeds 1067
1738	11s. 0d.: 14s. 0d.	5s. 9d.		Holkham Deeds
1739	14s. 6d.: 19s. 0d.	9s. 0d.		1067. Some of
1741	15s. 6d.			the figures are
1742	12s. 0d.	7s. 9d.		averages in-
1744	9s. 0d.	5s. 3d.		cluding inter-
1746		5s. 9d.		nal payments
1749	14s. 3d.: 15s. 0d.	6s. 6d.: 7s. 9d.		Posting Book,
1750	14s. 0d.: 16s. 6d.	6s. 9d.: 8s. 0d.		1749-54
1751		7s. 0d.: 7s. 3d.	7s. 0d.	(Muniment
1752	16s. 0d.: 17s. 0d.	7s. 0d.: 8s. 0d.		Room)
1753	13s. 0d.: 18s. 0d.	7s. 6d.: 8s. 6d.		
1754	12s. 0d.	6s. 0d.: 8s. 6d.	8s. 0d.	
1755	11s. 0d.	5s. 0d.: 5s. 6d.		Posting Book,
1756	18s. 0d.	6s. 6d.: 10s. 0d.	7s. 6d.: 10s. 0d.	1755-9
1757	24s. 0d.	9s. 0d.: 11s. 0d.	12s. 3d.	(Muniment
1758		6s. 0d.: 9s. 0d.	7s. 0d.	Room)
1759	12s. 0d.: 13s. 0d.	5s. 6d.: 6s. 6d.		
1759	13s. 0d.: 13s. 6d.	6s. 0d.: 6s. 6d.	7s. 0d.	Country account
1760	13s. 0d.: 14s. 0d.	5s. 6d.: 7s. 0d.		book
1761	12s. 6d.	6s. 6d.: 8s. 0d.		(Muniment
1762	16s. 0d.	8s. 4d.: 10s. 0d.	10s. 6d.: 11s. 3d.	Room)
1763	15s. 0d.	7s. 6d.: 9s. 6d.	11s. 0d.	
1764	19s. 0d.: 25s. 0d.	8s. 6d.: 9s. 6d.	10s. 0d.	
1765		8s. 4d.: 11s. 0d.	11s. 3d.	
1766	24s. 6d.	10s. 6d.: 14s. 0d.		
1767	22s. 4d.	9s. 0d.: 12s. 6d.		
1781	23s. 6d.: 25s. 0d.			Corn book
1782	19s. 0d.: 20s. 0d.			(Game Larder)
1783	26s. 6d.: 27s. 0d.	12s. 0d.: 13s. 6d.		
1784		12s. 0d.: 15s. 0d.		
1785		8s. 0d.: 12s. 0d.		
1786		10s. 0d.: 11s. 0d.		
1787		10s. 0d.: 11s. 0d.		

Appendix 5

	Wheat	Barley	Rye	
1815	26s. 6d.: 33s. 0d.	11s. 0d.: 15s. 0d.		General receipts and payments (Game Larder)
1816		13s. 0d.: 26s. 0d.		Farm accounts,
1817	35s. 0d.: 63s. 0d.	13s. 0d.: 32s. 0d.		1817–36
1821	20s. 0d.: 34s. 0d.	6s. 0d.: 18s. 0d.		(Game Larder)
1822	17s. 0d.: 23s. 6d.	7s. 0d.: 15s. 0d.		
1826	26s. 0d.: 35s. 0d.	15s. 0d.: 19s. 6d.		

Livestock prices were also important—especially those for the produce of sheep. Wool made an increasingly valuable contribution to the Norfolk farmers' economy, as the following figures show: the prices are in shillings per tod.

1724	6s. 0d.		1726	6s. 0d.
1725	5s. 3d.		1727	6s. 0d.
			1728	6s. 0d.
1732	10s. 0d.		1734	7s. 0d.
1733	11s. 0d.		1735	7s. 0d.
			1736	7s. 0d.
1749	10s. 0d.		1754	10s. 0d.
1750	13s. 0d.		1755	11s. 0d.
1751	13s. 0d.		1756	12s. 6d.
1752	12s. 0d.		1757	13s. 6d.
1753	9s. 6d.		1758	14s. 0d.
1760	15s. 6d.		1764	17s. 6d.
1761	13s. 0d.		1765	16s. 0d.
1762	12s. 6d.			
1763	15s. 0d.		1767	17s. 0d.
1817	57s. 0d.		1818	80s. 0d.
1823	46s. 0d.		1824	56s. 0d.

(References are as for corn figures for the same years.)

Appendix 5

Otherwise it is difficult to compare prices for sheep from one period to another, especially since most of the sheep sold for immediate slaughter were internal sales. However, prices received for lambs give an indication of the way the value of sheep was moving:

1724	3s. 8d.	1732	4s. 0d.
1725	6s. 0d.	1733	4s. 0d.
		1734	4s. 6d.
1727	3s. 0d.	1735	4s. 9d.
1752	5s. 6d.	1753	6s. 0d.
1761	7s. 0d.	1766	7s. 0d.
1763	6s. 0d.	1767	8s. 6d.
1764	7s. 0d.		
1819	28s. 0d.	1822	11s. 0d. to 14s. 0d.
1821	20s. 0d.		

(Same references.)

As for sheep sold for slaughter, the assumed internal price seems to have moved from 3s. 6d. or so a stone in the 1720s and 1730s to 4s. 1d. in the 1750s and 1760s.

Finally, evidence on prices for bullocks is unsatisfactory: the assumed internal price moved from 3s. 6d. per stone in the 1730s to 4s. 0d. in the 1760s.

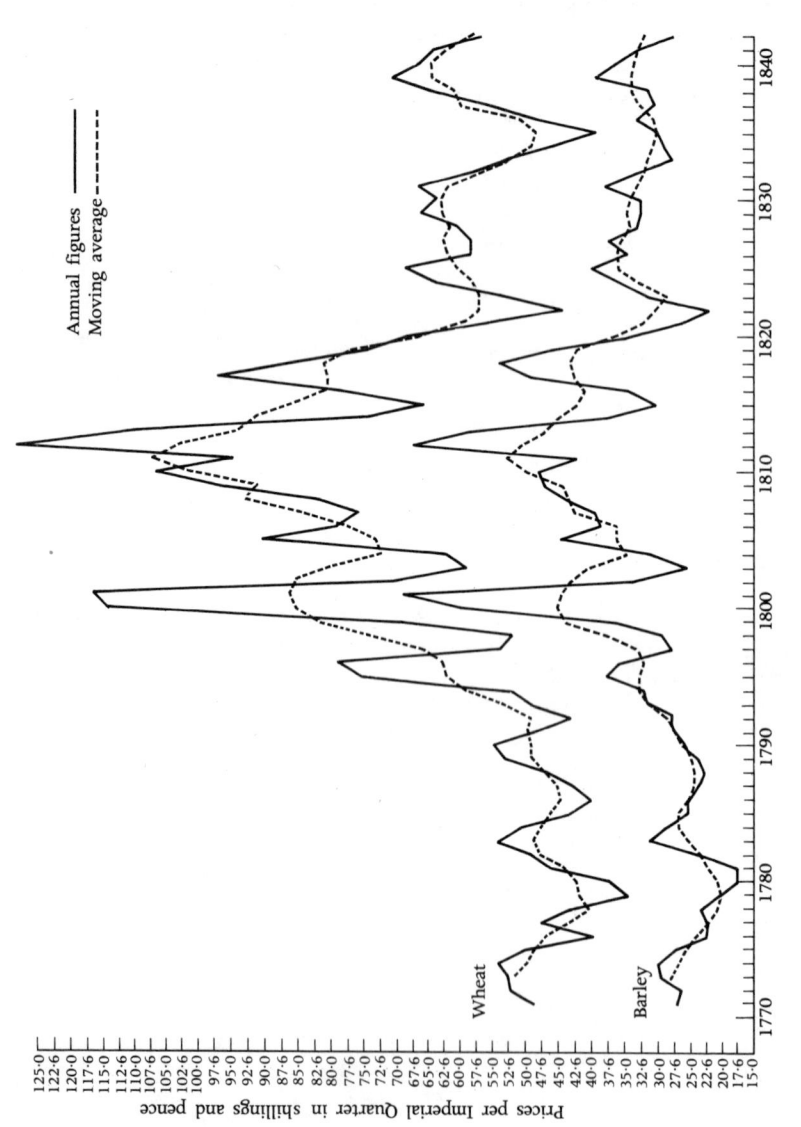

National average prices for wheat and barley 1771–1842

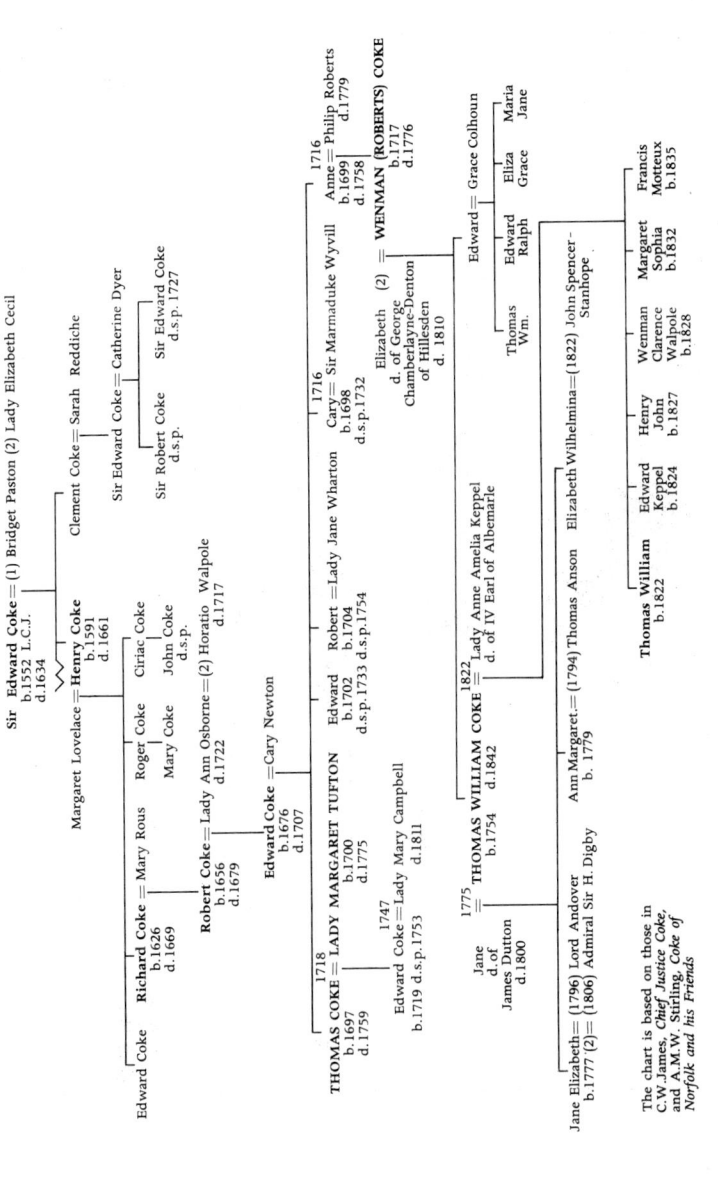

CHART OF DESCENT

Sir **Edward Coke** = (1) Bridget Paston (2) Lady Elizabeth Cecil
b.1552 L.C.J.
d.1634

Children:
- Margaret Lovelace = **Henry Coke** b.1591 d.1661
- Clement Coke = Sarah Reddiche
- Sir Edward Coke = Catherine Dyer

Henry Coke children:
- Roger Coke = Mary Coke
- Ciriac Coke
- John Coke d.s.p.
- Sir Robert Coke d.s.p.
- Sir Edward Coke d.s.p.1727

Robert Coke = Lady Ann Osborne = (2) Horatio Walpole
b.1656 d.1679 / d.1722 / d.1717

- Edward Coke
- **Richard Coke** = Mary Rous b.1626 d.1669
- **Edward Coke** = Cary Newton b.1676 d.1707

1718
THOMAS COKE = **LADY MARGARET TUFTON**
b.1697 d.1759 / b.1700 d.1775

1747
Edward Coke = Lady Mary Campbell
b.1719 d.s.p.1753 / d.1811

1775
Jane = **THOMAS WILLIAM COKE**
d. of James Dutton / b.1754 d.1842
d.1800

1716
Cary = Sir Marmaduke Wyvill
b.1698 d.s.p.1732

1716
Anne = Philip Roberts
b.1699 d.1779
d.1758

WENMAN (ROBERTS) COKE
b.1717 d.1776

Elizabeth (2) = d. of George Chamberlayne-Denton of Hillesden d.1810

- Edward = Grace Colhoun
  - Edward Ralph
  - Thomas Wm.
- Eliza Grace
- Maria Jane

Ann Margaret = (1794) Thomas Anson
b.1779

Elizabeth Wilhelmina = (1822) John Spencer-Stanhope

1822
Thomas William = Lady Anne Amelia Keppel
b.1822 / d. of IV Earl of Albemarle

- Edward Keppel b.1824
- Henry John b.1827
- Wenman Clarence Walpole b.1828
- Margaret Sophia b.1832
- Francis Motteux b.1835

Jane Elizabeth = (1796) Lord Andover
b.1777 (2) = (1806) Admiral Sir H. Digby

The chart is based on those in C.W. James, *Chief Justice Coke*, and A.M.W. Stirling, *Coke of Norfolk and his Friends*.

The holders of the Holkham estates are in bold type; those holding the estate within the period 1707–1842 are in capitals.

213

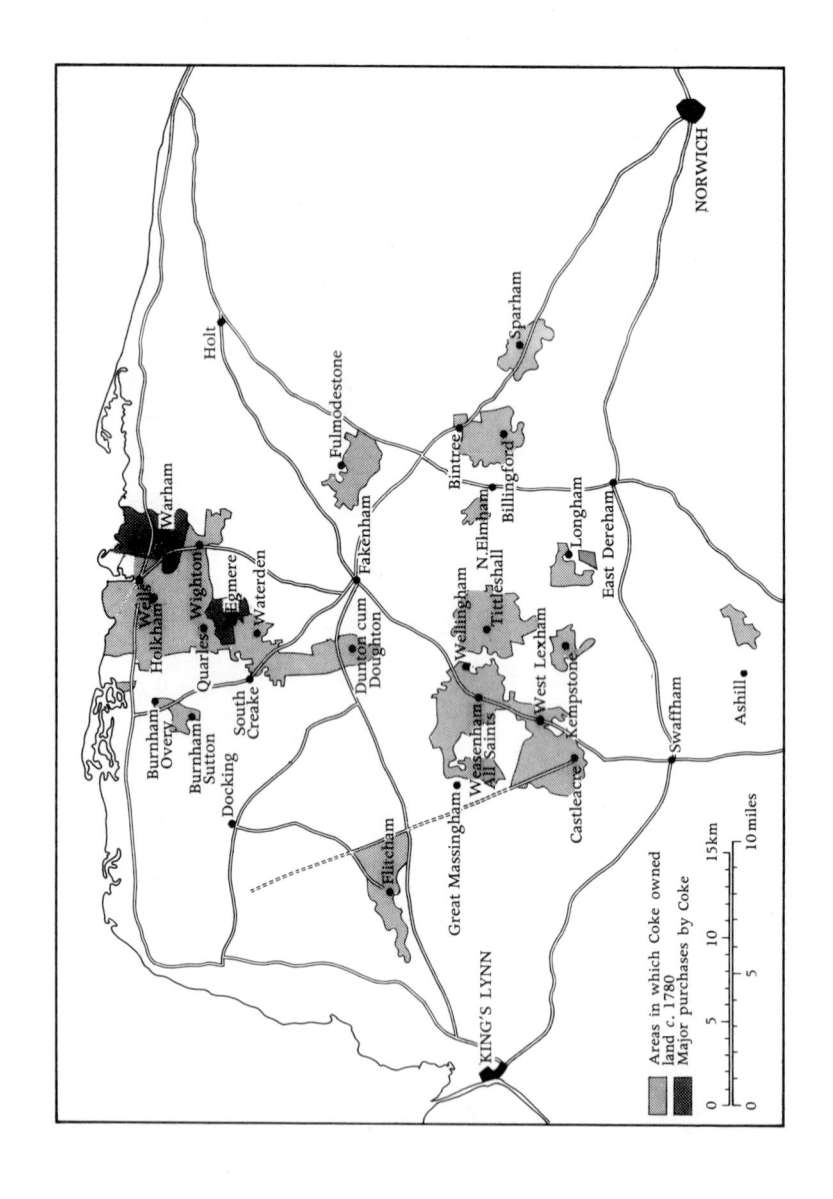
Holt
Fulmodestone
Sparham
NORWICH
Warham
Wells
Holkham
Wighton
Egmere
Waterden
Quarles
Fakenham
Dunton cum Doughton
Bintree
Billingford
N. Elmham
Tittleshall
Wellingham
Longham
East Dereham
West Lexham
Kempstone
Weasenham All Saints
South Creake
Burnham Overy
Burnham Sutton
Docking
Great Massingham
Castleacre
Swaffham
Ashill
Flitcham
KING'S LYNN
Areas in which Coke owned land *c.* 1780
Major purchases by Coke
0 5 10 15km
0 5 10 miles

INDEX

Abbott, Henry (tenant), 145
Abell, Thomas (tenant), 45
accounting methods, 6, 22
Agecroft turnpike trust, 197
Agricultural Returns, 1901 (Cd. 1121), 207
Almack, B., *On the Agriculture of Norfolk*, 160
Anderson, Francis (tenant), 45
Applegate, Robert (tenant), 100
Appleyard, George, 34, 57 n., 62
Argyll, Jane (Warburton), duchess of, 26, 69
Argyll, John Campbell, 2nd duke of, 26, 69
arrears of rent (1736), 40; (1814-24), 146; (1821-2), 148-9; (1834), 153
Ashill, Norfolk, 88
Ashworth, J., 175-8

Bacon, R. N., 75, 116 n., 122 n., 135 n., 136 n., 145, 171 n.
Baker, William, 94, 135, 154-5, 167-8, 187
Bakewell, Robert, 115, 119-20
Banks, Sir Joseph, 117, 120 n.
Barclays Bank Ltd., 134 n., 198 n.
barley, 11, 50, 52, 57-8, 68, 73, 104-13, 124, 136, 139-40, 149, 158, 162-3, 172
Bawdeswell, Norfolk, 96, 99
Bawdeswick workhouse, 197
Beauchamp, lady, 193, 196
Bedford, Francis Russell, 5th duke of, 116
Bedford, John Russell, 6th duke of, 117
Beeston, Robert (tenant), 111-12
beet, 172-3
Bertie, Charles, 1
Bevis Marks, London, 36, 83, 91, 126
Biedermann, H. A., 88
Billingford, Norfolk, 4, 83, 85-6, 88, 96, 99, 157
 Beckhall farm, 32
Binney, Thomas, 32
Bintree, Norfolk, 83, 88
Blaikie, Francis, 63, 66, 97, 100-1, 103 n., 121, 127, 175, 186
 corrects Coke myths, 75-9
 on soils of estate, 83
 becomes steward (1816), 135
 writes on agriculture, 135-6
 advises tenants, 136
 his energetic management, 136-7
 reports on estates and drafts forms for leases, 137
 personality, 137-8
 retires to Scotland, fears democracy, 138
 his leases, 138-47
 presses tenants to pay and predicts ruin (1822), 149
 attacks Castlereagh, 150
 deals with hard-pressed tenants, 151-2
 complains of excessive farm buildings, 154
 claims drainage of grassland more valuable than improving arable, 155 n.
 no vacant farms, 156
 on the four-course rotation and its dangers, 157-8
 explains use of oil cake as fertilizer, 159
 advocates large farms to enable sheep folding, 164
 helps to suppress riot (1831), 165-6
 writes that labour-saving machinery going out of use, 171
 complains of Coke's carelessness and goes to Manchester to find coal, 176
 denounces tenant, leases coal, 177
 manages lighthouse, 180-1
 works to extend its tenure, 182-3
 sets out Coke's financial position (1822), 188-92
 hopes Coke will see its seriousness, 191
 agrees on sale of land and hopes Coke will not fritter away proceeds, 192
 his salary cut, 195
 pleased by financial recovery but fears future, 196
 clears debts, 197-8
 enforces husbandry covenants, 199
Blomfield, John (tenant), 144
Blyth, Mr., 159
Blyth, William (tenant), 64, 106
borrowing, 1, 12-13, 17, 20, 23, 27, 61-2, 131-4, 200
 types of lenders, 29-34, 133-4, 195
Boys, Thomas, 114-15, 120
Branford, J. B. (tenant), 111
Braybroke, Richard Griffin Neville, 3rd baron, 183-4, 187

Index

break, meaning of, 204
Brett, Thomas (tenant), 74–6
Brettingham, M., *Plans, Elevations and Sections of Holkham*, 24 n.
Brougham, Samuel, 66, 135
Buckingham, Katherine, duchess of, 31
Burlington, Richard Boyle, earl of, 24
Burnhams, Norfolk, 27, 86
  Burnham Overy, 86, 165
  Burnham Sutton, 86–7, 159
  Burnham Thorpe, 165
Bylaugh, Norfolk, 10

Caird, James, *English Agriculture in 1850–1*, 145, 159, 171
Campbell, Duncan, 91
capital, accumulation and movements of:
  movement away from estates, 1822–42, 197–8
  movement into the estates, 134, 200
  see borrowing, repayment, improvements, investment, land purchases and sales, repairs, saving
Carr, John, of Massingham, 7, 38, 41
Carr, John, the younger, 41–2, 45, 50, 54–5, 64, 73, 104
Castleacre, Norfolk, 4, 11, 43, 45, 83, 87, 89, 90, 96, 101, 151, 157, 159–60
  Abbey farm, 45, 152
  Wicken farm, 41, 84, 145, 162
cattle, 58, 115, 139, 173, 211
  black cattle, 52
  Devons, 117–18, 121–2, 173
  Norfolks, 118–19, 122
  Scots cattle, 122
  Shorthorns, 122
  Herefords, 122
Cauldwell, Ralph, 27 n., 28 n., 34, 39, 56, 61–9, 73 n., 84, 100–1, 104, 135
Chamberlen, Hugh, 15, 17–20, 28
Chaplin, W. R., 36 n.
Charles II, king, 35, 182
Child's, bankers, 62
Christ's College, Cambridge, 38
Clayton, Sir John, 35
Clerke, George, *The Landed Man's Assistant*, 6 n.
clover, 8, 10, 49–52, 55, 68, 73, 102–3, 105, 139, 158, 172
coal, 175–8
Cobb, Benjamin, 180
Cocks, Charles, 61
cocksfoot, 139, 157, 172

Coke, Ann Margaret (1779–1843), da. of Coke of Norfolk, 130
Coke, Anne (1699–1758), 2, 61
Coke, Lady Anne Amelia (Keppel), Countess of Leicester (1803–44), 2nd wife to Coke of Norfolk, 130, 188
Coke, Cary (Newton) (1680–1707), 1
Coke, Cary (1698–1732), 2
Coke, Clement (1594–1629), 70
Coke, Sir Edward (d. 1727), 1, 126
Coke, Sir Edward (Lord Chief Justice) (1552–1634), 70
Coke, Edward (1676–1707), 1
Coke, Edward (1702–33), 2, 31, 70
Coke, Edward (1719–53) (son to Thomas Coke, 1st earl of Leicester), 25–6, 38, 128
Coke, Edward (d. 1837), brother to Coke of Norfolk, 70, 126, 197
Coke, Elizabeth, Mrs. (Denton) (d. 1810), 126, 128
Coke, Elizabeth Wilhelmina (1795–1873) (da. of Coke of Norfolk), 130, 138, 170, 189
Coke, Henry (1591–1661), 70
Coke, Jane Elizabeth (b. 1777) (da. of Coke of Norfolk), 130
Coke, John (of Baggrave, Leics.), a guardian, 1, 14
Coke, Lady Mary (Campbell) (1727–1811), 26, 29, 38, 69, 128
Coke, Margaret (Tufton), countess of Leicester, *see* Tufton, Margaret
Coke, Robert (1704–54), 2
Coke, Thomas (Sir Thomas Coke (1725), Baron Lovell (1728), earl of Leicester (1744)) (1697–1759), 1, 2
  comes of age and marries, 12
  South Sea Bubble and, 12–20
  financial position in 1722–33, 21–3
  joint postmaster-general, 23
  pawns plate, 23
  works with Kent and Burlington, 23–4
  directs spending on building in his will and interested in income, 25
  sells land, 26
  tries to reduce debts, 27
  concentrates estate in Norfolk, 27, 37–8
  debts at death, 27
  personalty, 28, 62
  settlement of estates, 29
  borrows from stewards, 34
  lighthouse grant to, 35
  encourages improvement, 40–9, 56–7

Index

agricultural revolution in his lifetime but remembered as virtuoso, 60
death and will, 61
debt left by, 61-2
sinking fund to repay debt, 62, 69
trust reposed by in steward, 63
pleased to see farm, 73
debts descend to Coke of Norfolk who disregards will of, 128, 190
restrictive effect of will, 151
his lighthouse, 202-3
Coke, Thomas William, 'Coke of Norfolk', 1st earl of Leicester of Holkham (1754-1842)
fame as landlord aspect of romantic movement, 60
succeeds to estates (1776), 61
attacks steward, 64
gives way, 65
takes over Wenman's estates, 69-70
myths about, 69, 71-82
devotes estates to support brother, 70
Coke responsible for myths, 78-82
his political views, 78
spends more on buildings, less on improvement of soil than great uncle, 94
boasts of spending on improvements, 94
gains from enclosure of commons, 99
objects to successive straw crops, 103
self-congratulation on his leases, 112
little evidence on extent of interference with tenants, 113
personality makes his farm famous, 114
welcomes visitors, 115
begins sheep-shearings, 116
raises social status of farmers, 118
changes in sheep in Norfolk, 119-22
advocates Devon cattle, 122-3
on drilling, 123
dependent on rents and income in 1776, 127
fails to carry out great-uncle's will, 128-9
spends more than income, 129
spending on daughters and politics, 130-1
debts in 1822, 131
has bank overdraft and no investments outside estates, 134
orders steward to report on estates, 137
improvements, more tenants' function than under Coke's predecessors, 142
calls for reduction in public spending and denounces ministers, 148
keeps farms in particular families, 149
good reputation as landlord, 150
charity to tenants, 151-2
defends government, 153
boasts of spending on improvement, 154
his spending mainly on buildings, 155
declines to buy North Creake, 156
recommends four-course rotation on good land, 158
advocates gypsum and inquires about nitrates, 160
denounces R.A.S.E., agricultural chemistry, and Sir H. Davy but praises Scottish agriculture, 161
suppresses riot and buys bloodhounds, 165-6
welcomes New Poor Law, 167
supports emigration, 167-8
ceases support, 168
boasts of crops in 1832, 172
congratulated on his bullocks by butchers, boasts of his sheep and imports pigs from Naples, 173
fails to understand potential value of coal, 175
casual attitude to money, 176
loses income, 178
apologizes for defects in his lighthouse, 180
tries to retain it, 182-3
his ownership petitioned against, 184
zeal for political reform unaffected by self-interest, 185
denounces Tories, praises C. J. Fox and boasts of his own agricultural achievements, 185-6
refuses to sell lighthouse, 187
accepts offer, 188
his financial crisis, 188-93
relief at paying daughter's portion and fear of repeal of corn laws, 189
takes crisis lightly, 191
his financial recovery, 194-8
gives reasons for refusing North Creake, 195
dies (1842), 197
leaves small personal estate, but rich real estate, 197-8
his impact on agricultural development exaggerated and that of his predecessors understated, 199
encouraged tenants by preaching and example, 200

Coke, Thomas William (b. 1793), Coke of Norfolk's nephew, 128, 197
Coke, Wenman (Roberts) (1717–76), 28, 61, 63–4, 69, 70, 128, 175
Coke estates:
  location, 1, 37, 83, 126–7
  value (1707), 1, 84; (1720–30), 22–3; (1718–50), 37–8; (1776–1815), 126–7; (1821–42), 194
  spending on, priority of, 22, 195
  Norfolk, concentration in and consequent losses, 27, 93, 178
  soils of, 83–4
  ownership pattern in, 86–8
  total crops grown on, 1790–7, 113
  farms on, no vacancies in 1831, 156, 167
  see enclosure, improvements, investment, marling, rents, size of farms
coleseed, 139
Collison, William (tenant), 31
Complete Farmer, The, 53
Cooper, widow (tenant), 137
cottages, 168
Coutts, bankers, 198 n.
Cowper, Nathaniel, 31
Coxsell, Charles, 202
Crick, Francis, 135
Croft, A. D., 193
Cross Keys Wash Bridge, 195

Darby, H. C. and Saltmarsh, J., *The Infield-Outfield System*, 204 n.
Darcy, Mr., 14, 16–17, 19
Dark, farrier, 33
Defoe, D., 9 n.
Dewing, Thomas (tenant), 145
Dewing, William, 42, 49
Dickson, P. G. M., *Financial Revolution in England*, 13 n., 15 n., 20 n.
Donyat, Dorset, 27
Down & Co., 90
drainage, 142, 155
drilling, 116–17, 119, 123–4, 158, 172, 200
Drosier, John, 42
Drury Lane theatre, 195, 197
Dugmore, J., surveyor, 88, 93
Dungeness lighthouse, 23, 34–6, 126, 178–89, 193, 196–7, 202–3
Dunton, Norfolk, 27, 31, 65, 73, 87, 89, 157
Durweston, Dorset, 27
Dutton, James, 1st Baron Sherborne, 66

Egmere, Norfolk, 92, 126, 164
elections, 79, 130–1
Elliott, John (bricklayer), 33
Elliott, John (tenant), 42
Elliott, William, 32
Ellis, K., *Post Office*, 23 n.
Ellman, John, of Glynde, 114, 120, 121
Ellman of Shoreham, 120
Elmham, Norfolk, 83
Emerson, Stephen, 135
emigration of labourers, 167–8
enclosure, 5, 40–52, 73, 84–8, 96–7, 99, 134, 156, 200
enclosure Acts, 43, 88, 96, 99, 100
estates, Coke, *see* Coke estates

family settlements and provisions:
  jointures, 2, 22, 26, 37, 62, 69 n., 126, 128
  marriage portions and daughters, 2, 12, 23, 26, 34, 130, 189
  younger brothers and sons, 2, 69–70, 190
  settlements, 29, 38, 91
  will of 1st Lord Leicester, 61, 128–9, 132, 188, 190
Farquhar, John, 192
Fellows, William, 28, 69
fertilizers, 136, 139, 156–60, 165, 171, 173, 200
Finch, Daniel Lord, 15, 17–20, 30–1
Fitzroy, General the Hon. William (tenant), 121, 123–4, 137, 150, 155–7, 159
Fitzwilliam estates, 66
Flitcham, Norfolk, 31, 50, 83, 87, 90, 102
  Abbey farm, 46, 48–9, 102
  Hall farm, 48
  Harpley Dam farm (New farm), 48–50, 107–8, 112, 155, 168
  Little Appleton farm, 48
Fodder, William (tenant), 75
Forby, Mitchell (tenant), 104
Foxley, Norfolk, 96, 99
Franklin, John (tenant), 48
Fulmodestone, Norfolk, 4, 7 n., 10, 27, 42–3, 53, 83, 88, 96, 99
Fussell, G. E., 72 n.

Gardiner, Richard, 64
Garwood, John (tenant), 160
Gentleman's Magazine, 51
Gibbs, E. H. (tenant), 117, 143
Gibson, Thomas, 13 n., 23
Gold and Silver Company, 14, 15
Gordon, James, 91

Index

Goulburn, Henry, 184
government securities, 2, 198
Gower, Mary (Tufton), countess of (sister to Margaret, countess of Leicester) (1701-85), 63
grass, 8, 11, 49, 51, 55, 68, 73, 102-13, 139-40, 155 n., 157, 162-3, 205
Gurney & Co., bankers, 90, 134, 197-8

Habakkuk, H. J., 24 n.
Hammersley's, bankers, 198
Hanrott, P. A., 93, 138, 147, 182-3, 189-93, 197-8
Harden, John (lighthouse keeper), 202
Hargrave, H. (lighthouse agent), 202
Hase, Mary, Mrs., 28, 69
Hastings, John (tenant), 97-9, 108
Hastings, Thomas (tenant), 85, 101
Hayes, Brian, 131 n.
Haylet, Thomas (senior) (tenant), 46
Haylet, Thomas (junior) (tenant), 46
Heagren, Edmund (tenant), 146, 152
Heard, William (tenant), 84-5, 89
Herbert, Miss, 198
Hervey, Lord, *Memoirs*, 24 n.
Hill, Charles (tenant), 137
Hill, Francis, 28
Hill, William (tenant), 107, 112
Hillesden, Bucks, 126-8, 192, 194, 197
Hillyard, C., *Practical Farming and Grazing*, 81
Holkham, 4, 10, 11, 29, 32, 56, 73-6, 86, 165-6
  Branthill Farm, 32, 42, 58, 161
  Hall farm, 57; *and see* Park farm
  Honclecronkdale farm, 9, 11, 74, 114
  Longlands farm, 41-2, 114
  Staithe farm, 114
  marshes, 59, 86
  Ostrich Inn, 74-5
Holkham Hall:
  building of, 22-4, 69
  cost of, 24-5, 39, 56, 62-3
Holkham sheep-shearings, 78, 80, 115-19, 121
  cattle at, 122-3
  drilling at, 123
  end of, 170
  guests at, 117-18
  prizes offered (1803), 116
  sheep, hiring and sales of (1806), 117; (1821), 169
  toasts drunk (1810), 117

Holland, corn export to, 51
Home farm, *see* Park farm
horses, 58, 115, 173
House of Commons, Select Committee on Foreign Trade, 1822, 3rd Report, 182
household expenses, 21-2, 189-90, 206
Huddleston, mercer, 33
Hudson, J. (tenant), 149, 159-60
Hume, Joseph, 118, 181, 184-8
Hunloke, Lady (Coke of Norfolk's sister), 133
Hutcheson, Archibald, *Calculations and Remarks*, 17 n.
Hyde (tenant), 175, 177

improvements, 7, 8, 13, 39-52, 54, 56-7, 67, 93-7, 134, 142, 153-5
  see investment, repairs
inclosure, *see* enclosure
interest, rates of (1675), 1 n.; (1716-25), 12; (1720), 13, 15, 17; (1721-30), 23; (1741-65), 28; (1700-60), 30-1, 33, 49; (1822), 132; (1775-1822), 133; (1822-36), 192-3
investment, 7, 13, 14, 40-2, 48-9, 51, 56-7, 93-5, 97, 134, 153-6, 195, 197-8, 200
  none outside estates, 134, 197
  see improvements, land purchases and sales, repairs
Ives, Clement (tenant), 49

Jackson, Richard, 89
James, C. W., *Chief Justice Coke* . . ., 23 n., 25 n., 31 n., 77
Jickling, N. (lighthouse agent), 180
jointures, *see* family settlements and provisions
Jones, Daniel, 32
Jourdain, M., *William Kent*, 24 n.

Keary, H. W., 48 n., 84, 158, 162-4, 168, 171
Keith, Major James, 51 n.
Kempstone, Norfolk, 83-4, 88-9, 121, 123-4, 144, 155
Kent, Claridge & Co. (valuers), 101
Kent, Nathaniel, 8 n., 54, 77-8, 101-3, 105
Kent, William (architect), 23-4
Kent, William (tenant), 42
Kettlestone, Norfolk, 100, 168
Kindersley (drainage constructor), 13
King, Edward, 33
Kingsdown, Kent, 83, 91, 126
King's Lynn theatre, 197

Index

Knatts, Henry, 31
Knightley, Staffs., 27
Knowles, Andrew, 177–8

labourers, 59, 60, 165–8, 171, 205
Lamb, Sir Matthew, 27–8, 31, 34, 61–2, 69
Lamb, Peniston, 13, 14, 20, 29, 30
Lamerie, Paul, 13
Lancashire estates, 70, 90, 126, 132, 175–8
land purchases and sales, 4, 25–7, 38–9, 49, 66–7, 84–5, 89, 90–3, 126, 129–30, 156, 192, 195
  policy of purchasing, 38–9, 86 n., 89, 93, 134, 200
Lanes, John (lighthouse keeper), 202
La Rochefoucauld, François de, 66
Leak, B. (surveyor), 99 n.
leases, 5, 40, 45, 53–4, 62, 67, 100–1, 103–4, 137–47, 151–2, 155
  husbandry covenants in, 45, 49, 50, 54–5, 67–8, 73, 95–6, 102–4, 106, 111–12, 137–47, 199
  prohibition of successive white straw crops in, 103–6, 109, 111–13, 139–41, 143–4
Lee, William, 41–2
Leeds City Library, 3 n.
Leeds, Thomas (tenant), 169
Leeds, Thomas Osborne, 1st duke of, 2 n.
Limekiln, 58
Liverpool, Robert Jenkinson, 2nd earl of, 182
London Gazette, 207
London Insurance, 14, 15
London University, 195
Longham, Norfolk, 43, 53, 83, 85, 88, 96–7
Longham Hall farm, 97–9, 101–2, 108–9, 205
lucerne, 57, 139
Lydd, Kent, 36, 180–1, 184

McGregor, O. R., 72 n.
Macpherson, D., *Annals of Commerce*, 18 n.
Mallett, Benoni, 31, 65, 68, 73, 89, 100
Manchester, 175
mangel wurzel, 117, 119, 158, 164, 200
manure, *see* fertilizers
Marchand, J., 66 n.
marling, 7, 8, 40–2, 48, 50–2, 54, 57, 73, 94, 97, 142, 155
marriage settlements, *see* family settlements and provisions

Marshall, William, *Rural Economy of Norfolk*, 100, 204 n.
Martin, Sir Mordaunt, 119
Mason, William (tenant), 112
Massingham, Norfolk, 7, 9, 33, 41, 45, 50, 64, 68, 73, 83, 85, 87, 105–7, 109–11
Master, Thomas, 66
Middleton, near Stockport, Cheshire, 176
Mileham, Norfolk, 43 n., 168
Miller, Philip, *Gardener's Dictionary*, 52–3
Minster Lovell, Oxon., 83, 126, 195
Money, William (tenant), 74
Mordant, John, *The Complete Steward*, 54–5
mortgages, 1, 2, 13, 17, 20, 25–30, 61, 63, 90, 132–3, 193, 195–6, 200
Mosby, J. E. G., *Norfolk*, 163 n.
Murrell (farmer), 7

Newcastle, Thomas Pelham-Holles, 1st duke of, 25 n.
New Poor Law, 167
Newton, Mr., 186–7
Newton, Sir John, 1, 12
nonsuch, 57
Norfolk Chronicle, 117 n., 118 n., 120 n., 121 n., 122 n., 187
North Creake, Norfolk, 156
North Elmham, Norfolk, 9, 88
Norwich, Bishop of, 90
Norwich Literary Institution, 195
Norwich Mercury, 75–6, 78 n., 80 n., 116 n., 117 n., 118 n., 120 n., 148 n., 186 n.
Norwich Union Fire Office, 195

oats, 50, 52, 58, 68, 107–13, 124, 140, 144, 162–3, 205
open field, 40, 42, 51–2, 84–8, 99
  see enclosure
Overman, H. J. (tenant), 137, 145, 197
Overman, J. R. (tenant), 87, 159
Owen, Robert, 118

Park farm, Holkham (Home farm or Hall farm), 57–60, 68–9, 71, 74–6, 84, 90, 114, 119–25, 160, 169–74, 200
  crops grown on and yields (1782–7), 124–5; (1826, 1832), 172
  extent of (1780–1816), 114; (1823–42), 170
  fertilizers on, 171
  profit and loss on (1817–26), 170
  pupils on, 170–1

Index

stock on, nature and value of (1816), 114; (1817–23), 169; (1817, 1826), 172; (1842), 173
  uneconomic nature of, 170, 200
  see Holkham sheep-shearings
Pattern, Michael, 32
peas, 50, 103–13, 124, 139–40, 173
Peel, Sir Robert, 185
Pelham, Henry, 31
Petingale, George, 143
pigs, 58, 115–16
  Neapolitans, 173
  Suffolks, 173
Plumb, J. H., 9, 10 n.
Popplewell, Cdr. Matthew, R.N., 179
Portbury, Somerset, 83, 91, 126
Portman, Henry, 17
prices, 39, 51, 52, 96, 99, 145, 149, 169, 205, 207–11
Prothero, R. E., Lord Ernle, 68, 71–2
Purdy, Thomas (tenant), 149, 157
Pyndar, Philip, 61
Pyndar, William, 61

Quarles, Norfolk, 38, 87, 117, 157, 170

railways, 167
rape, rapeseed, rapecake, 139, 158–9
rates, poor, 75, 149, 165, 205
Rebow, Sir Isaac, 36 n.
Reddish, Lancs., 175
Reeve, John (tenant), 117–18, 137
Reform Club, 197
rents (in 17th cent. and 1706–17), 1–5; (1720–30), 22–3; (1730–59), 26–7; (1718–59), 37–42, 45, 53, 66, 89; (1758–76), 67; (1776–1816), 76–7, 95–6, 126–7; (1814–24), 146; (1820–42), 152–3, 189–90, 194
  coal rents, 177
  effects of enclosure on, 45–6, 48, 50, 89
  estate underrented, 147
  rebates in rent (1821–3), 150, 153
  see arrears of rent
  repairs, 7, 56–7, 67, 94–5, 134, 141, 149, 153–4, 189
  see improvements, investment
repayment, 1, 26–8, 62–3, 131, 193, 195–6
Rhoades, Thomas (tenant), 89
Riches, N., *Agricultural Revolution in Norfolk*, 73 n., 77–8, 104
rick-burning, 166

Rigby, E., 71, 73, 75 n., 76, 77 n., 79, 80
riot (1831), 165–6
Rix, John (tenant), 144
Roberts, Philip, 12, 28, 61
Roberts, Wenman, *see* Coke, Wenman
Robinson, F. J., 183
Rogers, J. E. Thorold, *History of Agriculture and Prices*, 3 n., 57 n.
rotations of crops, 8, 50–2, 55, 68, 101–13, 138–45, 159
  four-course rotation, 68, 102, 104–7, 138–9, 141–2, 145, 156–9, 162–4, 171, 174, 199–200
Rudd, Isaac (tenant), 144
Ryburgh, Norfolk, 96
rye, 50, 52, 57–8, 68, 73, 111, 124

sainfoin, 139
Sanctuary, Thomas (tenant), 86, 102, 106–7, 112, 133, 145
Sanders, H. G., *British Crop Husbandry*, 104
Saunders, H. W., 9 n., 10
Savage, Henry, 135
saving, 1
Savory, Henry, 31
Scott, W. R., *Joint-Stock Companies*, 13 n., 14 n., 16
Sharpe, Jeremiah (tenant), 74
sheep, 8, 40–3, 50–2, 57–9, 115, 159, 164–5, 169, 200, 210–11
  Leicesters, 116–17, 119–20, 122 n.
  Southdowns, 116–17, 119–22
  Norfolks, 119–20
  Merinos, 120
  Hampshires, 173
sheep-shearings, *see* Holkham sheep-shearings
Shepherd, Henry (tenant), 152
Shillington, Dorset, 27
Shipman, Elizabeth, 35
Shipp, William (tenant), 151
Sinclair, Sir John, 77, 123, 159
size of farms, 53, 89, 164
Skipton, Edmond (tenant), 53
Smith of Dishley, 120 n.
Smith, Humphrey, 7, 54, 57
Snow, Thomas, 14–15, 17–20
South Creake, Norfolk, 42, 87
South Sea Stock, 13–20
Sparham, Norfolk, 4, 7, 83, 88, 90, 96, 99, 103 n., 112
Spencer, George John, 2nd earl, 156, 195
Spencer, John Charles, 3rd earl, 79, 80

stewards, 4, 7, 33-4, 57, 61-9, 135
  see Appleyard, George; Baker, William; Blaikie, Francis; Brougham, Samuel; Cauldwell, Ralph; Crick, Francis; Gardiner, Richard; Kent, Nathaniel; Smith, Humphrey; Wyatt, William
Stibbard, Norfolk, 96
Stirling, A. M. W., *Coke of Norfolk*, 74
  Letter Bag of Lady Elizabeth Spencer-Stanhope, 138 n., 170, 189
Stokes, Henry, 136
Stokes, Miss M. V., 198 n.
Styleman, Nicholas, 64
subletting, 7
Sweden, demand for corn in, 51

Tann, Thomas (tenant), 32, 74
Tattersall, George, 169
Taunton, W. E., 195 n.
taxation, 2, 3, 59, 127-8, 148-9, 205-6
Terry, Stephen (lighthouse supervisor), 181
Thomson, Poulett, 185
threshing machines, 165, 171-2
Thurston, Dr., 28, 69
tithes, 59
Tittleshall (and Godwick), Norfolk, 4, 31, 38, 43, 46, 83-4, 88, 102, 104, 157, 204
  Godwick farm, 103, 111
Tollet, George, 120
Tooke, T., *History of Prices*, 125 n., 169 n.
Townshend estates, 9, 10, 72 n., 89, 104
Trinity House, 35, 179-83, 186-8
Tufton, Margaret, baroness Clifford, countess of Leicester (1700-75)
  marries Thomas Coke (1718), 12, 29
  brings lighthouse, 34
  life-tenant of the estates (1770-75), 28, 61
  likes steward and dislikes heir, 63-5
  purchases no land, 67
Tufton, Mary (Mrs.), 28, 132
Tufton, Richard, 5th earl of Thanet, 35, 182
Tufton, Sackville, 15, 17-20
Tufton, Thomas, 6th earl of Thanet, 12, 23, 34
Turner, Sir John, 91
Turner, John (tenant), 151

turnips, 8-10, 49-53, 55, 57, 68, 102, 104-5, 138-40, 142 n., 158, 162-3, 172-3, 205
Swedish turnips, 116, 164, 172

Venn, J. A., *Foundations of Agricultural Economics*, 141 n.
vetches, 50, 103-13, 173

wages, 59
Waller, Edmund, 20, 23, 31
Waller, Edmund (valuer), 101
Walpole, Lady Anne, 2, 22, 37
Walpole estates, 9, 10
Walsingham, Norfolk, 166
Warham, Norfolk, 80, 90-2, 96, 126, 144
water meadows, 116-17
Waterden, Norfolk, 43, 55, 87, 103, 107, 112, 204
Weasenham, Norfolk, 4, 8, 10, 42, 83, 86-7, 89-90, 96, 99, 102, 106-7, 112, 145, 155, 157
Wellingham, Norfolk, 83, 87, 89, 96, 99
Wellington, Arthur Wellesley, 1st duke of, 185
Wells, Norfolk, 96, 156
Western, Charles Callis, 1st baron, 161
West Lexham, Norfolk, 83, 87, 90, 160
wheat, 11, 50, 57-8, 68, 73, 102, 105-13, 143, 149, 158-60, 162-4, 172-3, 205
Whitaker, Mr., 91
Whormby, J. (lighthouse agent), 202
Wiffin, William (tenant), 146, 152
Wightman, Thomas (tenant), 85
Wighton, Norfolk, 27-8, 38, 87, 90, 102, 111-12, 117, 152, 157
  Crab Castle farm, 146, 152
Wilbraham, Roger, 133
Winn, Henry (tenant), 74, 75
Worlidge, John, 9
Wyatt, Samuel, 179
Wyatt, William, 101
Wyvill, Sir Marmaduke, 12

Yarmouth, William Paston, earl of, 4
yields of corn, 80 n., 97, 160, 205
Yorke, Charles, 28, 30, 61
Young, Arthur, 72 n., 73, 87, 101, 104-6, 112, 114-16, 119, 120 n., 122-3, 142 n., 159 n.